GLOBAL STUDIES

THE WORLD AT A GLANCE

Aster Tessema

THE WORLD AT A GLANCE

Staff

Larry Loeppke	*Managing Editor*
Nichole Altman	*Development Editor*
Maggie Lytle	*Cover*
Kari Voss	*Typesetting Supervisor/Co-designer*
Jean Smith	*Typesetter*
Sandy Wile	*Typesetter*
Karen Spring	*Typesetter*
Mary Ejlali	*Pager/Designer*

Copyright

Cataloging in Publication Data
I. Title: The World at a Glance. Tessema, Aster.
ISBN 0-07-340408X Library of Congress Reg. # 1-223-801

First Edition

Printed in the United States of America 234567890QPDQPD98765

**THIS BOOK IS DEDICATED
TO ALL THE CHILDREN OF THE WORLD
WHO I HOPE WILL MAKE THIS EARTH
AND FUTURE LIVEABLE PLANETS
MUCH BETTER PLACES TO LIVE**

PREFACE

A "valuable book for home or office and a resource for school projects: social studies, geography, world cultures, etc."

It is useful to know some of the important facts about the world we live in (such as a nation's capital, type of government, currency, major languages, population, ethnic groups, religions, political structure, cultural background, climate, economics, industrial revolutions and contributions to civilization, etc.) Clearly, it is close to impossible to be an expert on all of these areas; the range of subject matters is far too great. One can, however, attempt to retain an understanding of some of these essential facts, in order to make useful applications. Therefore, this book is intended to assist in such a process.

As you know, there are many sources with different data out there, but I'm confident that I have used up-to-date and accurate information that I felt are from reliable sources. It should be noted, however, that some of the information might already have changed by the publication of this book.

Although, various sources were used in the compilation of this book, I wish especially to acknowledge the assistance of Embassies and Consulates here in Washington, D.C., for their exceptional cooperation in providing me with accurate facts and findings, as well as the following International Organizations and The U.S. Government Agencies: The United Nations Information Centres/Services, International Demographic Data Center, The World Bank, U.S. State Department, The Library of Congress, The U.S. Congress, U.S. Central Intelligence Agency, U.S. Bureau of Census, U.S. Geological Association and National Aeronautic Space Administration (NASA).

My heartfelt gratitude goes out to American General Supplies, Inc., not only for sponsoring the previous editions of this book, but also for promoting it globally.

I would also like to express my special thanks to Nichols C. Adamopoulos, P.E., Alphatec, P.C., Eric Baldwin, Tilahun Bekele, The United Nations (ECA), Clayborne E. Chavers, Sr., Esq., The Chavers Law Firm, P.C., L. Paul Clausen, Bernard Demczuk, Ph.D., George Washington University, Sharon L. Donohoe, The American Institute of Architects, James W. Giles, AIA, Raquel Jordan, Raymond & Gilda Lambert, Nicole Maharaj, The United States Conference of Mayors, Shirley Papp and Jonathan Rich, who encouraged me in my work and without these wonderful people the book would never have been written. To David F. Anderson, III, P.E., Alphatec, P.C., Thomas H. Franks, Professor of Geography, University of Virginia and Dr. Vinod Thomas, The World Bank, who read the first draft and supplied valuable comments and suggestions.

Responsibility for the accuracy of the materials in this book rests, of course, with me alone.

CONTENTS

The Nine Planets and World Statistics

THE NINE PLANETS

- **Mercury** (closest to the Sun)
- **Venus**
- **EARTH**
- **Mars**
- **Jupiter** (largest)
- **Saturn**
- **Uranus**
- **Neptune**
- **Pluto** (smallest & furthest from the Sun)

METRIC CONVERSION

Km (kilometers) **to Miles**	(km x **0.6214**= Miles)
Miles **to Km**	(miles x **1.6090** = Km)
Sq Km **to Sq Miles**	(sq km x **0.3861** = Sq Miles)
Sq Miles **to Sq Km**	(sq miles x **0.25900** = Sq Km)
Feet **to Meters**	(ft x **0.3048** = Meters)
Meters **to Feet**	(meters x **3.2810** = Ft)

EARTH'S DIMENSIONS

Estimated Age 4.6 Billion Years
Current Population 6,446,000,000
Type of Water (97% **Salt**), (3% **Fresh**)

(57,393,000 sq miles)	**Land Area**	(148,647,000 sq km)	**29.1%**
(139,544,000 sq miles)	**Total Water Area**	(361,419,000 sq km)	**70.9%**
(12,944,000 sq miles)	**Ocean Area**	(335,258,000 sq km)	
(196,938,000 sq miles)	**Surface Area**	(510,064,000 sq km)	
(92,956,000 miles)	**Distance from the Sun / Average**	(149,566,000 km)	
(91,400,000 miles)	**Distance from the Sun / Closest**	(147,063,000 km)	
(94,500,000 miles)	**Distance from the Sun / Farthest**	(152,051,000 km)	
(24,901 miles)	**Circumference at the Equator**	(40,066 km)	
(24,851 miles)	**Circumference at the Poles**	(39,992 km)	
(7,925 miles)	**Diameter at the Equator**	(12,753 km)	
(7,898 miles)	**Diameter at the Poles**	(12,710 km)	
(3,963 miles)	**Radius at the Equator**	(6,376 km)	
(3,949 miles)	**Radius at the Poles**	(6,355 km)	

Orbit Speed/ The Earth orbits the Sun at (66,629 mph), (107,229 km per hour)
Orbit Frequency/ The Earth orbits the Sun every **365** days, **5** hours, **48** minutes and **46** seconds

CONTINENTS
Africa
Antarctica
Asia
Europe
North America
South America
Oceania

CONTINENTS POPULATION	
Asia	(3,935,000,000)
Africa	(891,000,000)
Europe	(705,000,000)
North America	(483,000,000)
South America	(342,000,000)
Oceania	(31,000,000)
Antarctica	(- 0 -)

CONTINENTS (By Land Mass)	
Asia	(17,128,500 sq miles) (44,362,815 sq km)
Africa	(11,707,000 sq miles) (30,321,130 sq km)
North America	(9,363,000 sq miles) (24,250,170 sq km)
South America	(6,875,000 sq miles) (17,806,250 sq km)
Antarctica	(5,500,000 sq miles) (14,245,000 sq km)
Europe	(4,057,000 sq miles) (10,507,630 sq km)
Oceania	(2,966,136 sq miles) (7,682,292 sq km)

INDEPENDENT NATIONS (By Continent)	
Africa	(53)
Asia	(47)
Europe	(43)
North America	(23)
Oceania	(14)
South America	(12)
Antarctica	(0)
TOTAL:	192

OCEAN FLOORS
Antarctic Ocean
Arctic Ocean
Atlantic Ocean
Indian Ocean
Pacific Ocean

POLAR REGIONS
(North Pole)
Arctic, Arctic Circle,
Arctic Ocean
(South Pole)
Antarctica, Antarctic Circle
Antarctic Ocean or
(Southern Ocean)

LONGEST RIVERS

Nile, Africa	(4,145 miles/ 6,671 km)
Amazon, South America	(3,915 miles/ 6,300 km)
Chang Jiang (Yangtze), Asia	(3,900 miles/ 6,276 km)
Mississippi, North America	(3,741 miles/ 6,019 km)
Ob'lrtysh-black lrtysh, Europe	(3,362 miles/ 5,411 km)
Yenisey-Angara, Europe	(3,100 miles/ 4,989 km)
Huang He (Yellow), Asia	(2,877 miles/ 4,630 km)
Amur, Asia	(2,744 miles/ 4,416 km)
Lena, Europe	(2,734 miles/ 4,399 km)
Congo/*Zaire*, Africa	(2,718 miles/ 4,374 km)
Mackenzie, North America	(2,635 miles/ 4,241 km)
Mekong, Asia	(2,610 miles/ 4,200 km)
Niger, Africa	(2,566 miles/ 4,129km)

SMALLEST NATIONS (Under 100,000 Population)

Vatican City/**Holy See**	921
Tuvalu	11,636
Nauru	13,048
Palau	20,303
San Marino	28,880
Monaco	32,409
Liechtenstein	33,717
St. Kitts & Nevis	38,958
Marshall Islands	59,071
Antigua & Barbuda	68,722
Dominica	69,029
Andorra	70,549
Seychelles	81,188
Grenada	89,502

LARGEST NATIONS (By Land Mass)

Russia	17,075,383 sq km (6,592,812 sq mil)
Canada	9,976,128 sq km (3,851,787 sq mil)
China	9,559,690 sq km (3,691,000 sq mil)
United States	9,384,658 sq km (3,623,420 sq mil)
Brazil	8,506,663 sq km (3,284,426 sq mil)
Australia	7,682,292 sq km (2,966,136 sq mil)
India	3,287,588 sq km (1,269,339 sq mil)
Argentina	2,776,661 sq km (1,072,070 sq mil)
Kazakhstan	2,715,097 sq km (1,048,300 sq mil)

LARGEST NATURAL LAKES

Caspian Sea	(370,999 sq km) 143,243 sq miles
Lake Superior, USA-Canada	(82,414 sq km) 31,820 sq miles
Lake Victoria, Tanzania-Uganda	(69,215 sq km) 26,724 sq miles
Lake Huron, USA-Canada	(59,596 sq km) 23,010 sq miles
Lake Michigan, USA	(58,016 sq km) 22,400 sq miles
Lake Aral, Kazakhstan-Uzbekistan	(41,000 sq km) 15,830 sq miles
Lake Tanganyika, Tanzania-Congo	(32,764 sq km) 12,650 sq miles
Lake Baykal, Russia	(31,500sq km) 12,162 sq miles
Lake Great Bear, Canada	(31,328 sq km) 12,096 sq miles
Lake Nyasa, Malawi-Mozambique-Tanzania	(29,928 sq km) 11,555 sq miles
Lake Great Slave, Canada	(28,570 sq km) 11,031 sq miles
Lake Erie, USA-Canada	(25,754 sq km) 9,940 sq miles
Lake Winnipeg, Canada	(24,390 sq km) 9,417 sq miles
Lake Ontario, USA-Canada	(19,529 sq km) 7,540 sq miles
Lake Ladoga, Russia	(18,399 sq km) 7,104 sq miles
Lake Balkhash, Kazakhstan	(18,200 sq km) 7,027 sq miles

MAJOR SEAS

Caribbean Sea	(970,000 sq mil/ 2,512,300 sq km)
Mediterranean Sea	(969,000 sq mil/ 2,509,710 sq km)
South China Sea	(895,000 sq mil/ 2,318,050 sq km)
Bering Sea	(875,000 sq mil/ 2,266,250 sq km)
Gulf of Mexico	(600,000 sq mil/ 1,554,000 sq km)
Sea of Okhotsk	(590,000 sq mil/ 1,528,100 sq km)
Arabian Sea	(578,504 sq mil/ 1,498,325 sq km)
East China Sea	(482,000 sq mil/ 1,248,380 sq km)
Yellow Sea	(480,000 sq mil/ 1,243,200 sq km)
Sea of Japan	(389,000 sq mil/ 1,007,510 sq km)
Hudson Bay	(317,500 sq mil/ 822,325 sq km)
North Sea	(222,000 sq mil/ 574,980 sq km)
Black Sea	(185,000 sq mil/ 479,150 sq km)
Red Sea	(169,000 sq mil/ 437, 710 sq km)
Baltic Sea	(163,000 sq mil/ 422,170 sq km)

SURFACE POINTS

Above / Below Sea Level

Highest Point
Mt. Everest, Asia
29,035 feet / 8,850 meters

Lowest Point
Dead Sea, Asia
1,349 feet / -411 meters

TALLEST MOUNTAINS

Mount Everest, Nepal-China	(29,034 ft/ 8,850 meters)
Godwin Austen (K2), Pakistan/Kashmir-China	(28,250 ft/ 8,611 meters)
Kanchenjunga, Nepal-India	(28,169 ft/ 8,586 meters)
Lhotse, Nepal-China	(27,940 ft/ 8,516 meters)
Makalu I, Nepal-China	(27,766 ft/ 8,463 meters)
Cho Oyu, Nepal-China	(26,906 ft / 8201 meters)
Dhaulagiri, Nepal	(26,795 ft/ 8,167 meters)
Manaslu I, Nepal	(26,781 ft/ 8,163 meters)
Gasherbrum, Pakistan-China	(26,740 ft/ 8,150 meters)
Nanga Parbat, Pakistan (Kashmir)	(26,660 ft/ 8,126 meters)
Annapurna I, Nepal	(26,545 ft/ 8,091 meters)

THE WORLD'S MAJOR RELIGIONS

Religion	Share
Muslims	**20%**
Catholics	**17%**
Hindus	**13%**
Buddhists	**6%**
Protestants	**6%**
Orthodox	**3%**
Atheists	**2%**
Anglicans	**1%**
Sikhs	**0.39%**
Jews	**0.23%**

LARGEST DESERTS

Sahara, Northwestern Africa	3,500,000 sq miles/ 9,065,000 sq km
Arabian, Southwestern Asia	1,000,000 sq miles/ 2,590,000 sq km
Gobi, Mongolia & Northeastern China	500,000 sq miles/ 1,295,000 sq km
Gibson, Western & Southern Australia	250,000 sq miles/ 647,500 sq km
Great Victoria, Western/Central Australia	250,000 sq miles/ 647,500 sq km
Kalahari, Southwestern Africa	190,000 sq miles/ 492,100 sq km

LARGEST ISLANDS

Area (sq km)	Island	Area (sq miles)
(2,175,600 sq km)	**Greenland, North America**	(840,000 sq miles)
(789,950 sq km)	**New Guinea, Oceania**	(305,000 sq miles)
(751,100 sq km)	**Borneo, SE Malaysia**	(290,000 sq miles)
(586,376 sq km)	**Madagascar, Africa**	(226,400 sq miles)
(507,454 sq km)	**Baffin, Canada**	(195,928 sq miles)
(424,760 sq km)	**Sumatra, Indonesia**	(164,000 sq miles)
(227,920 sq km)	**Honshu, Japan**	(88,000 sq miles)
(218,896 sq km)	**Great Britain, Europe**	(84,400 sq miles)
(217,290 sq km)	**Victoria, Canada**	(83,896 sq miles)
(196,236 sq km)	**Ellesmere, Canada**	(75,767 sq miles)
(189,034 sq km)	**Celebes, Indonesia**	(72,986 sq miles)
(151,238 sq km)	**So. Island, New Zealand**	(58,393 sq miles)
(126,501 sq km)	**Java, Indonesia**	(48,842 sq miles)
(114,444 sq km)	**No. Island, New Zealand**	(44,187 sq miles)
(108,860 sq km)	**Newfoundland, Canada**	(42,031 sq miles)
(104,981 sq km)	**Cuba, North America**	(40,533 sq miles)
(104,688 sq km)	**Luzon, Philippines**	(40,420 sq miles)
(103,000 sq km)	**Iceland, Europe**	(39,768 sq miles)

SMALLEST NATIONS
(Under 1,000 Sq. Miles)

Vatican City / **Holy See**	0.17 (sq miles) 0.44 sq km
Monaco	0.75 (sq miles) 2 sq km
Nauru	8 (sq mil) 21 sq km
Tuvalu	10 (sq mil) 26 sq km
San Marino	24 (sq mil) 62 sq km
Liechtenstein	62 (sq mil) 161 sq km
Marshall Islands	70 (sq mil) 181 sq km
Saint Kitts & Nevis	103 (sq mil) 269 sq km
Maldives	116 (sq mil) 300 sq km
Malta	122 (sq mil) 316 sq km
Grenada	133 (sq mil) 344 sq km
Saint Vincent & The Grenadines	150 (sq mil) 388 sq km
Barbados	166 (sq mil) 430 sq km
Antigua & Barbuda	171 (sq mil) 443 sq km
Seychelles	176 (sq mil) 455 sq km
Andorra	188 (sq mil) 487 sq km
Palau	188 (sq mil) 487 sq km
Singapore	226 (sq mil) 585 sq km
Saint Lucia	238 (sq mil) 616 sq km
Bahrain	240 (sq mil) 622 sq km
Tonga	270 (sq mil) 699 sq km
Micronesia	271 (sq mil) 702 sq km
Dominica	290 (sq mil) 751 sq km
Kiribati	291 (sq mil) 754 sq km
Sao Tome & Principe	372 (sq mil) 963 sq km
Comoros	719 (sq mil) 1,862 sq km
Mauritius	790 (sq mil) 2,046 sq km
Luxembourg	999 (sq mil) 2,587 sq km

LARGEST CITIES
(8 Millions/+ Population)

City	Population
Mumbai/Bombay, India	(10)
Sao Paulo, Brazil	(10)
Seoul, South Korea	(10)
Jakarta, Indonesia	(9)
Karachi, Pakistan	(9)
Istanbul, Turkey	(8)
Mexico City, Mexico	(8)
Moscow, Russia	(8)
New York City, USA	(8)
Shanghai, China	(8)
Tokyo, Japan	(8)

Note: Numbers shown are the population within the recognized city limits, and do not include people living in the immediate surrounding area outside of the established border of the city.

THE WORLD'S MAJOR LANGUAGES

Language	Speakers
Chinese (*Mandarin*)	950 million
English	500 million
Hindi	490 million
Spanish	375 million
Russian	280 million
Arabic	250 million
Bengali	220 million
Portuguese	190 million
Japanese	130 million
German	121 million
Korean	98 million
French	85 million
Chinese (*Wu*)	82 million
Javanese	80 million
Chinese (*Yue*)	79 million

The Independent Nations of the World

COUNTRY *(CAPITAL CITY)* *(CONTINENT)*	CURRENCY	PREDOMINANT RELIGIONS	OFFICIAL / MAJOR LANGUAGES
A			
Afghanistan (*Kabul*) *(AS)*	Afghani	Sunni & Shi'a Muslim	Pashtu/Dari
Albania (*Tirana*) *(EU)*	Lek	Muslim/Orthodox/Catholic	Albanian/Tosk
Algeria (*Algiers*) *(AF)*	Algerian Dinar	Sunni Muslim	Arabic
Andorra (*Andorra la Vella*) *(EU)*	Euro	Catholic	Catalan
Angola (*Luanda*) *(AF)*	Kwanza	(*) Catholic/Protestant	Portuguese
Antigua & Barbuda (*St.John's*) *(NA)*	East Caribbean Dollar	Protestant/Catholic	English
Argentina (*Buenos Aires*) *(SA)*	Argentine Peso	Catholic	Spanish
Armenia (*Yerevan*) *(AS)*	Dram	Protestant (Apostolic)	Armenian
Australia (*Canberra*) *(OC)*	Australian Dollar	Protestant/Catholic	English
Austria (*Vienna*) *(EU)*	Euro	Catholic	German
Azerbaijan (*Baku*) *(AS)*	Azerbaijani Manat	Muslim/Orthodox	Azeri/Russian/Armenian
B			
Bahamas, The (*Nassau*) *(NA)*	Bahamian Dollar	Protestant/Catholic	English/Creole
Bahrain (*Manama*) *(AS)*	Bahraini Dinar	Sunni & Shi'a Muslim	Arabic
Bangladesh (*Dhaka*) *(AS)*	Taka	Muslim/Hindu	Bengali

Africa *(AF)* **Asia** *(AS)* **Europe** *(EU)* **North America** *(NA)* **South America** *(SA)* **Oceania** *(OC)*
Former names are in italic: ***(Burma/*** Myanmar, ***Cambodia/*** Kampuchea, Congo, Democractic Republic of the/***Zaire,*** Cote d'Ivoire/ ***Ivory Coast,*** Samoa/ ***Western Samoa, Vatican City/*** Holy See*)*
Communaute Financiere Africaine (CFA) Franc
(*) *Asterisk denotes that many people practice native religions, based on the worship of various spirits*

COUNTRY *(CAPITAL CITY)* *(CONTINENT)*	CURRENCY	PREDOMINANT RELIGIONS	OFFICIAL / MAJOR LANGUAGES
Barbados *(Bridgetown) (NA)*	Barbadian Dollar	Protestant/Catholic	English
Belarus *(Minsk) (EU)*	Belarusian Ruble	Orthodox	Belarusian
Belgium *(Brussels) (EU)*	Euro	Catholic/Protestant	Dutch/French/German
Belize *(Belmopan) (NA)*	Belizean Dollar	Catholic/Protestant	English
Benin *(Porto-Novo) (AF)*	CFA Franc	(*) Protestant/Catholic/Muslim	French
Bhutan *(Thimphu) (AS)*	Ngultrum&Indian Rupee	Buddhist/Hindu	Dzongkha
Bolivia *(Sucre) (SA)*	Boliviano	Catholic	Spanish/Quechua/Aymara
Bosnia & Herzegovina *(Sarajevo) (EU)*	Marka	Muslim/Orthodox/Catholic	Croatian/Serbian/Bosnian
Botswana *(Gaborone) (AF)*	Pula	(*) Protestant/Catholic	English
Brazil *(Brasilia) (SA)*	Real	Catholic	Portuguese
Brunei *(Bandar Seri Begawan)(AS)*	Bruneian Dollar	Muslim	Malay
Bulgaria *(Sofia) (EU)*	Lev	Orthodox	Bulgarian
Burkina Faso *(Ouagadougou) (AF)*	CFA Franc	(*) Muslim	French
Burma /**Myanmar** *(Yangon/Rangoon) (AS)*	Kyat	Buddhist	Burmese
Burundi *(Bujumbura) (AF)*	Burundi Franc	(*) Catholic	Kirundi/French

COUNTRY *(CAPITAL CITY) (CONTINENT)*	CURRENCY	PREDOMINANT RELIGIONS	OFFICIAL / MAJOR LANGUAGES
C			
Cambodia / **Kampuchea** *(Phnom Penh)(AS)*	Riel	Buddhist	Khmer
Cameroon *(Yaounde) (AF)*	CFA Franc	(*) Protestant/Catholic/Muslim	English/French
Canada *(Ottawa) (NA)*	Canadian Dollar	Catholic/Protestant	English/French
Cape Verde *(Praia) (AF)*	Cape Verdean Escudo	(*) Catholic	Portuguese/Crioulo
Central African Republic *(Bangui)(AF)*	CFA Franc	(*) Protestant/Catholic/Muslim	French
Chad *(N'Djamena) (AF)*	CFA Franc	Muslim/Protestant/Catholic	French/Arabic
Chile *(Santiago) (SA)*	Chilean Peso	Catholic/Protestant	Spanish
China *(Beijing) (AS)*	Yuan	Officially Atheist/Daoist	Chinese/Mandarin/Yue/Wu
Colombia *(Bogota) (SA)*	Colombian Peso	Catholic	Spanish
Comoros *(Moroni) (AF)*	Comoran Franc	Sunni Muslim	Arabic/French
Congo, Democratic Republic of the */Zaire (Kinshasa)(AF)*	Congolese Franc	Catholic/Protestant	French
Congo, Republic of the *(Brazzaville)(AF)*	CFA Franc	Protestant	French
Costa Rica *(San Jose) (NA)*	Costa Rican Colon	Catholic/Protestant	Spanish
Cote d'Ivoire*/Ivory Coast (Yamoussoukro/ Abidjan)(AF)*	CFA Franc	(*) Muslim/Protestant/Catholic	French

COUNTRY *(CAPITAL CITY)* *(CONTINENT)*	CURRENCY	PREDOMINANT RELIGIONS	OFFICIAL / MAJOR LANGUAGES
Croatia *(Zagreb) (EU)*	Kuna	Catholic	Croatian
Cuba *(Havana) (NA)*	Cuban Peso	Catholic	Spanish
Cyprus *(Nicosia) (AS)*	Cypriot Pound&Turkish Lira	Orthodox/Muslim	Greek/Turkish
Czech Republic *(Prague) (EU)*	Czech Koruna	Catholic	Czech
D			
Denmark *(Copenhagen) (EU)*	Danish Krone	Protestant	Danish/English/Faroese/ Greenlandic
Djibouti *(Djibouti) (AF)*	Djiboutian Franc	Muslim	French/Arabic
Dominica *(Roseau) (NA)*	East Caribbean Dollar	Catholic/Protestant	English
Dominican Republic *(Santo Domingo)(NA)*	Dominican Peso	Catholic	Spanish
E			
East Timor *(Dili) (AS)*	US Dollar	Catholic	Tetum/Portuguese
Ecuador *(Quito) (SA)*	US Dollar	Catholic	Spanish
Egypt *(Cairo) (AF)*	Egyptian Pound	Sunni Muslim	Arabic
El Salvador *(San Salvador) (NA)*	US Dollar	Catholic	Spanish
Equatorial Guinea *(Malabo) (AF)*	CFA Franc	(*) Catholic	Spanish/French
Eritrea *(Asmara) (AF)*	Nakfa	Orthodox/Muslim	Tigrinya

COUNTRY *(CAPITAL CITY) (CONTINENT)*	CURRENCY	PREDOMINANT RELIGIONS	OFFICIAL / MAJOR LANGUAGES
Estonia (*Tallinn*) *(EU)*	Estonian Kroon	Protestant/Orthodox/Catholic	Estonian
Ethiopia (*Addis Ababa*) *(AF)*	Birr	Orthodox/Muslim	Amharic
F			
Fiji (*Suva*) *(OC)*	Fijian Dollar	Protestant/Hindu	English
Finland (*Helsinki*) *(EU)*	Euro	Protestant	Finnish/Swedish
France (*Paris*) *(EU)*	Euro	Catholic	French
G			
Gabon (*Libreville*) *(AF)*	CFA Franc	Protestant/Catholic	French
Gambia, The (*Banjul*) *(AF)*	Dalasi	Muslim	English
Georgia (*T'bilisi*) *(AS)*	Lari	Orthodox	Georgian
Germany (*Berlin*) *(EU)*	Euro	Protestant/Catholic	German
Ghana (*Accra*) *(AF)*	Cedi	(*) Protestant/Catholic	English
Greece (*Athens*) *(EU)*	Euro	Orthodox	Greek
Grenada (*St. George's*) *(NA)*	East Caribbean Dollar	Catholic/Protestant	English
Guatemala (*Guatemala City*) *(NA)*	Quetzal & US Dollar	(*) Catholic/Protestant	Spanish/Amerindian
Guinea (*Conakry*) *(AF)*	Guinean Franc	Muslim	French

COUNTRY *(CAPITAL CITY)* *(CONTINENT)*	CURRENCY	PREDOMINANT RELIGIONS	OFFICIAL / MAJOR LANGUAGES
Guinea-Bissau (*Bissau*) *(AF)*	CFA Franc	(*) Muslim/Protestant	Portuguese
Guyana (*Georgetown*) *(SA)*	Guyanese Dollar	Protestant/Catholic/Hindu	English/Amerindian/Creole
H			
Haiti (*Port-au-Prince*) *(NA)*	Gourde	(*) Catholic/Protestant	French/Creole
Honduras (*Tegucigalpa*) *(NA)*	Lempira	Catholic	Spanish/Amerindian
Hungary (*Budapest*) *(EU)*	Forint	Catholic/Protestant	Hungarian
I			
Iceland (*Reykjavik*) *(EU)*	Icelandic Krona	Protestant	Icelandic/English/ Nordic/German
India (*New Delhi*) *(AS)*	Indian Rupee	Hindu/Muslim	English/Hindi/Urdu
Indonesia (*Jakarta*) *(AS)*	Indonesian Rupiah	Muslim	Bahasa Indonesia
Iran (*Tehran*) *(AS)*	Iranian Rial	Sunni & Shi'a Muslim	Persian/Turkic/Kurdish
Iraq (*Baghdad*) *(AS)*	New Iraqi Dinar	Sunni & Shi'a Muslim	Arabic/Kurdish
Ireland (*Dublin*) *(EU)*	Euro	Catholic	English
Israel (*Jerusalem*) *(AS)*	New Israeli Shekel	Jewish/Muslim	Hebrew
Italy (*Rome*) *(EU)*	Euro	Catholic	Italian

COUNTRY *(CAPITAL CITY)* *(CONTINENT)*	CURRENCY	PREDOMINANT RELIGIONS	OFFICIAL / MAJOR LANGUAGES
J			
Jamaica (*Kingston*) *(NA)*	Jamaican Dollar	(*) Protestant	English/patois English
Japan (*Tokyo*) *(AS)*	Yen	Shinto/Buddhist	Japanese
Jordan (*Amman*) *(AS)*	Jordanian Dinar	Sunni Muslim/Orthodox	Arabic
K			
Kazakhstan (*Astana*) *(AS)*	Tenge	Sunni Muslim/Orthodox	Kazakh/Russian
Kenya (*Nairobi*) *(AF)*	Kenyan Shilling	(*) Protestant/Catholic	English/Kiswahili
Kiribati (*Tarawa*) *(OC)*	Australian Dollar	Catholic/Protestant	I-Kiribati/English
Korea, North (*Pyongyang*) *(AS)*	North Korean Won	Buddhist/Confucianist	Korean
Korea, South (*Seoul*) *(AS)*	South Korean Won	Buddhist/Protestant/Catholic	Korean
Kuwait (*Kuwait City*) *(AS)*	Kuwaiti Dinar	Sunni & Shi'a Muslim	Arabic
Kyrgyzstan (*Bishkek*) *(AS)*	Kyrgyzstani Som	Muslim/Orthodox	Russian/Kyrgyz
L			
Laos (*Vientiane*) *(AS)*	Kip	Buddhist/Protestant/Catholic	Lao
Latvia (*Riga*) *(EU)*	Latvian Lat	Protestant/Catholic/Orthodox	Latvian
Lebanon (*Beirut*) *(AS)*	Lebanese Pound	Muslim/Orthodox/Protestant	Arabic

COUNTRY *(CAPITAL CITY)* *(CONTINENT)*	CURRENCY	PREDOMINANT RELIGIONS	OFFICIAL / MAJOR LANGUAGES
Lesotho *(Maseru) (AF)*	Loti & South African Rand	(*) Protestant/Catholic	English
Liberia *(Monrovia) (AF)*	Liberian Dollar	(*) Catholic/Protestant/Muslim	English
Libya *(Tripoli) (AF)*	Libyan Dinar	Sunni Muslim	Arabic/Italian
Liechtenstein *(Vaduz) (EU)*	Swiss Franc	Catholic	German
Lithuania *(Vilnius) (EU)*	Litas	Catholic	Lithuanian
Luxembourg *(Luxembourg) (EU)*	Euro	Catholic	Luxembourgish/German/ French
M			
Macedonia *(Skopje) (EU)*	Macedonian Denar	Orthodox/Muslim	Macedonian/Albanian
Madagascar *(Antananarivo) (AF)*	Malagasy Franc	(*) Protestant/Catholic	French/Malagasy
Malawi *(Lilongwe) (AF)*	Malawian Kwacha	Protestant/Catholic/Muslim	English/Chichewa
Malaysia *(Kuala Lumpur) (AS)*	Ringgit	Muslim/Hindu/Buddhist/Daoist	Bahasa Melayu
Maldives *(Male) (AS)*	Rufiyaa	Sunni Muslim	Maldivian Dhivehi/English
Mali *(Bamako) (AF)*	CFA Franc	Muslim	French/Bambara
Malta *(Valletta) (EU)*	Maltese Lira	Catholic	Maltese/English
Marshall Islands *(Majuro) (OC)*	US Dollar	Protestant	English/Marshallese
Mauritania *(Nouakchott) (AF)*	Ouguiya	Muslim	Arabic/Pulaar/Soninke/Wolof

COUNTRY *(CAPITAL CITY) (CONTINENT)*	CURRENCY	PREDOMINANT RELIGIONS	OFFICIAL / MAJOR LANGUAGES
Mauritius *(Port Louis) (AF)*	Mauritian Rupee	Hindu/Catholic	English/Creole/French
Mexico *(Mexico City) (NA)*	Mexican Peso	Catholic	Spanish
Micronesia *(Palikir) (OC)*	US Dollar	Catholic/Protestant	English
Moldova *(Chisinau) (EU)*	Moldovan Leu	Orthodox	Moldovan/Russian
Monaco *(Monaco) (EU)*	Euro	Catholic	French
Mongolia *(Ulan Bator/ Ulaanbaatar)(AS)*	Togrog / Tugrik	Buddhist	Khalkha Mongol
Morocco *(Rabat) (AF)*	Moroccan Dirham	Sunni Muslim	Arabic/French
Mozambique *(Maputo) (AF)*	Metical	(*) Protestant/Catholic/Muslim	Portuguese
N			
Namibia *(Windhoek) (AF)*	Namibian Dollar/So.African Rand	(*) Protestant/Catholic	English/Afrikaans/German
Nauru *(None) (OC)*	Australian Dollar	Protestant/Catholic	Nauruan
Nepal *(Kathmandu) (AS)*	Nepalese Rupee	Hindu	Nepali
Netherlands, The *(Amsterdam) (EU)*	Euro	Catholic/Protestant	Dutch/Frisian
New Zealand *(Wellington) (OC)*	New Zealand Dollar	Protestant/Catholic	English/Maori
Nicaragua *(Managua) (NA)*	Gold Cordoba	Catholic	Spanish
Niger *(Niamey) (AF)*	CFA Franc	Muslim	French

COUNTRY *(CAPITAL CITY) (CONTINENT)*	CURRENCY	PREDOMINANT RELIGIONS	OFFICIAL / MAJOR LANGUAGES
Nigeria *(Abuja) (AF)*	Naira	Muslim/Protestant/Catholic	English
Norway *(Oslo) (EU)*	Norwegian Krone	Protestant	Norwegian
O			
Oman *(Muscat) (AS)*	Omani Rial	Muslim	Arabic
P			
Pakistan *(Islamabad) (AS)*	Pakistani Rupee	Sunni & Shi'a Muslim	Urdu/ English
Palau *(Koror) (OC)*	US Dollar	(*) Catholic/Protestant	English/Palauan/Sonsoralese Tobi/Angaur/Japanese
Panama *(Panama City) (NA)*	Balboa & US Dollar	Catholic	Spanish
Papua New Guinea *(Port Moresby)(OC)*	Kina	(*) Protestant	English/Motu
Paraguay *(Asuncion) (SA)*	Guarani	Catholic	Spanish/Guarani
Peru *(Lima) (SA)*	Nuevo Sol	Catholic	Spanish/Quechua
Philippines *(Manila) (AS)*	Philippine Peso	Catholic	Filipino/English
Poland *(Warsaw) (EU)*	Zloty	Catholic	Polish
Portugal *(Lisbon) (EU)*	Euro	Catholic	Portuguese
Q			
Qatar *(Doha) (AS)*	Qatari Rial	Muslim	Arabic

COUNTRY *(CAPITAL CITY)* *(CONTINENT)*	CURRENCY	PREDOMINANT RELIGIONS	OFFICIAL / MAJOR LANGUAGES
R			
Romania (*Bucharest*) *(EU)*	Leu	Orthodox	Romanian
Russia (*Moscow*) *(EU)*	Russian Ruble	Orthodox/Muslim	Russian
Rwanda (*Kigali*) *(AF)*	Rwandan Franc	Catholic/Protestant	Kinyarwanda/French/English
S			
Saint Kitts & Nevis (*Basseterre*) *(NA)*	East Caribbean Dollar	Protestant/Catholic	English
Saint Lucia (*Castries*) *(NA)*	East Caribbean Dollar	Catholic	English
Saint Vincent & The Grenadines *(Kingstown)(NA)*	East Caribbean Dollar	Protestant/Catholic	English/French patois
Samoa /*Western Samoa* (*Apia*) *(OC)*	Tala	Protestant/Catholic	Samoan/English
San Marino (*San Marino*) *(EU)*	Euro	Catholic	Italian
Sao Tome & Principe (*Sao Tome*) *(AF)*	Dobra	Protestant/Catholic	Portuguese
Saudi Arabia (*Riyadh*) *(AS)*	Saudi Riyal	Muslim	Arabic
Senegal (*Dakar*) *(AF)*	CFA Franc	Muslim	French
Serbia & Montenegro (*Belgrade*) *(EU)*	New Yugoslav Dinar/ Euro	Orthodox/Muslim	Serbian/Albanian
Seychelles (*Victoria*) *(AF)*	Seychelles Rupee	Catholic	English/French

COUNTRY *(CAPITAL CITY)* *(CONTINENT)*	CURRENCY	PREDOMINANT RELIGIONS	OFFICIAL / MAJOR LANGUAGES
Sierra Leone (*Freetown*) *(AF)*	Leone	(*) Muslim	English/Mende/Temne/Krio
Singapore (*Singapore City*) *(AS)*	Singapore Dollar	Buddhist/Muslim	Chinese/Malay/Tamil/English
Slovakia (*Bratislava*) *(EU)*	Slovak Koruna	Catholic	Slovak
Slovenia (*Ljubljana*) *(EU)*	Tolar	Catholic	Slovenian/Serbo-Croatian
Solomon Islands (*Honiara*) *(OC)*	Solomon Islands Dollar	Protestant/Catholic	English/Melanesian pidgin
Somalia (*Mogadishu*) *(AF)*	Somali Shilling	Sunni Muslim	Somali
South Africa (*Pretoria*) *(AF)*	Rand	(*) Protestant/Catholic	Afrikaans/English/Ndebele/ Pedi/Sotho/Swazi/Tsonga/ Tswana/Venda/Xhosa/Zulu
Spain (*Madrid*) *(EU)*	Euro	Catholic	Castilian Spanish
Sri Lanka (*Colombo*) *(AS)*	Sri Lankan Rupee	Buddhist/Hindu	Sinhala/Tamil/English
Sudan (*Khartoum*) *(AF)*	Sudanese Dinar	(*) Sunni Muslim	Arabic
Suriname (*Paramaribo*) *(SA)*	Surinamese Guilder	Hindu/Protestant/Catholic/Muslim	Dutch
Swaziland (*Mbabane*) *(AF)*	Lilangeni	(*) Catholic/Protestant/Muslim	English/siSwati
Sweden (*Stockholm*) *(EU)*	Swedish Krona	Protestant	Swedish
Switzerland (*Bern*) *(EU)*	Swiss Franc	Catholic/Protestant	German/French/Italian/ Romansch
Syria (*Damascus*) *(AS)*	Syrian Pound	Sunni Muslim	Arabic

COUNTRY *(CAPITAL CITY)* *(CONTINENT)*	CURRENCY	PREDOMINANT RELIGIONS	OFFICIAL / MAJOR LANGUAGES
T			
Tajikistan *(Dushanbe) (AS)*	Somoni	Sunni Muslim	Tajik/Russian
Tanzania *(Dodoma/ Dar es Salaam) (AF)*	Tanzanian Shilling	(*) Muslim/Protestant/Catholic	Swahili/English
Thailand *(Bangkok) (AS)*	Baht	Buddhist	Thai/English
Togo *(Lome) (AF)*	CFA Franc	(*) Protestant/Catholic/Muslim	French
Tonga *(Nuku'alofa) (OC)*	Pa'anga	(*) Protestant	Tongan/English
Trinidad&Tobago *(Port-of-Spain) (NA)*	Trinidad & Tobago Dollar	Catholic/Hindu/Protestant	English
Tunisia *(Tunis) (AF)*	Tunisian Dinar	Muslim	Arabic/French
Turkey *(Ankara) (AS)*	Turkish Lira	Sunni Muslim	Turkish
Turkmenistan *(Ashgabat) (AS)*	Turkmen Manat	Muslim	Turkmen/Russian/Uzbek
Tuvalu *(Funafuti) (OC)*	Australian/Tuvaluan Dollar	Protestant	Tuvaluan/English/ Samoan/Kiribati
U			
Uganda *(Kampala) (AF)*	Ugandan Shilling	(*) Catholic/Protestant/Muslim	English
Ukraine *(Kiev) (EU)*	Hryvnia	Orthodox	Ukrainian/Russian/Romanian/ Polish/Hungarian
United Arab Emirates *(Abu Dhabi) (AS)*	Emirati Dirham	Muslim	Arabic
United Kingdom *(London) (EU)*	British Pound	Protestant/Catholic	English/Welsh/Scottish

COUNTRY *(CAPITAL CITY)* *(CONTINENT)*	CURRENCY	PREDOMINANT RELIGIONS	OFFICIAL / MAJOR LANGUAGES
United States (*Washington D.C.*) *(NA)*	US Dollar	Protestant/Catholic	English
Uruguay (*Montevideo*) *(SA)*	Uruguayan Peso	Catholic	Spanish/Portunol
Uzbekistan (*Tashkent*) *(AS)*	Uzbekistani Sum	Muslim	Uzbek/Russian/Tajik
V			
Vanuatu (*Port-Vila*) *(OC)*	Vatu	(*) Protestant/Catholic	English/French/Bislama
Vatican City /**Holy See** (*None*) *(EU)*	Euro	Catholic	Italian/Latin/French
Venezuela (*Caracas*) *(SA)*	Bolivar	Catholic	Spanish
Vietnam (*Hanoi*) *(AS)*	Dong	(*) Buddhist/Daoist/Catholic	Vietnamese
Y			
Yemen (*Sanaa*) *(AS)*	Yemeni Rial	Muslim	Arabic
Z			
Zambia (*Lusaka*) *(AF)*	Zambian Kwacha	Catholic/Hindu/Muslim	English
Zimbabwe (*Harare*) *(AF)*	Zimbabwean Dollar	(*) Protestant/Catholic	English

Independent Nations by Continent

THE 53 INDEPENDENT NATIONS OF AFRICA
FLAG, POPULATION & LAND AREA (SQ. MILES / SQ. KILOMETERS)

ALGERIA
Pop: 32,531,853
Sq.Mil: 919,591
Sq.Km: 2,381,740

ANGOLA

Pop: 11,190,786
Sq.Mil: 481,351
Sq.Km: 1,246,700

BENIN
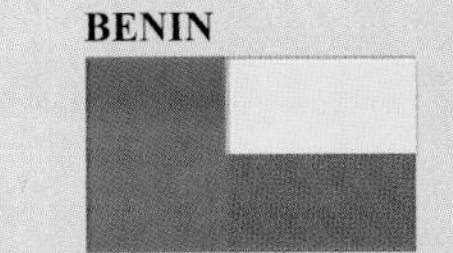
Pop: 7,460,025
Sq.Mil: 43,483
Sq.Km: 112,620

BOTSWANA
Pop: 1,640,115
Sq.Mil: 224,764
Sq.Km: 582,139

BURKINA FASO

Pop: 13,925,313
Sq.Mil: 105,869
Sq.Km: 274,200

BURUNDI

Pop: 6,370,609
Sq.Mil: 10,747
Sq.Km: 27,835

CAMEROON

Pop: 16,380,005
Sq.Mil: 183,568
Sq.Km: 475,441

CAPE VERDE

Pop: 418,224
Sq.Mil: 1,557
Sq.Km: 4,033

CENTRAL AFRICAN REP.

Pop: 3,799,897
Sq.Mil: 242,000
Sq.Km: 626,780

CHAD
Pop: 9,826,419
Sq.Mil: 495,752
Sq.Km: 1,283,998

COMOROS
Pop: 671,247
Sq.Mil: 719
Sq.Km: 1,862

CONGO, DEMO. REP. OF THE

Pop: 60,085,804
Sq.Mil: 905,063
Sq.Km: 2,344,133

CONGO, REP. OF THE

Pop: 3,039,126
Sq.Mil: 132,046
Sq.Km: 342,000

COTE D'IVOIRE/ *IVORY COAST*

Pop: 17,298,040
Sq.Mil: 124,504
Sq.Km: 322,465

DJIBOUTI

Pop: 476,703
Sq.Mil: 8,880
Sq.Km: 23,000

EGYPT

Pop: 77,505,756
Sq.Mil: 386,659
Sq.Km: 1,001,447

EQUATORIAL GUINEA

Pop: 535,881
Sq.Mil: 10,831
Sq.Km: 28,052

ERITREA

Pop: 4,561,599
Sq.Mil: 46,842
Sq.Km: 121,320

ETHIOPIA

Pop: 73,053,286
Sq.Mil: 471,776
Sq.Km: 1,221,900

GABON

Pop: 1,389,201
Sq.Mil: 103,346
Sq.Km: 267,666

GAMBIA, THE
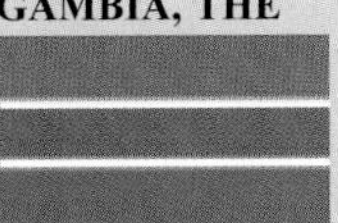
Pop: 1,593,256
Sq.Mil: 4,127
Sq.Km: 10,689

GHANA

Pop: 21,029,853
Sq.Mil: 92,099
Sq.Km: 238,536

GUINEA

Pop: 9,467,866
Sq.Mil: 94,925
Sq.Km: 245,856

GUINEA-BISSAU

Pop: 1,416,027
Sq.Mil: 13,948
Sq.Km: 36,125

KENYA
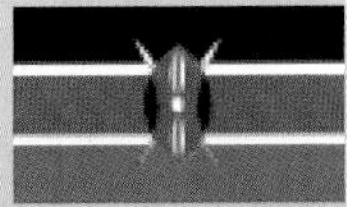
Pop: 33,829,590
Sq.Mil: 224,960
Sq.Km: 582,646

LESOTHO

Pop: 1,867,035
Sq.Mil: 11,720
Sq.Km: 30,355

LIBERIA

Pop: 3,482,211
Sq.Mil: 43,000
Sq.Km: 111,370

LIBYA

Pop: 5,765,563
Sq.Mil: 679,358
Sq.Km: 1,759,537

MADAGASCAR

Pop: 18,040,341
Sq.Mil: 226,657
Sq.Km: 587,041

MALAWI

Pop: 12,158,924
Sq.Mil: 45,747
Sq.Km: 118,485

MALI
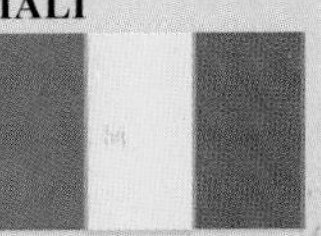
Pop: 12,291,529
Sq.Mil: 464,873
Sq.Km: 1,204,021

MAURITANIA

Pop: 3,086,859
Sq.Mil: 419,229
Sq.Km: 1,085,803

MAURITIUS
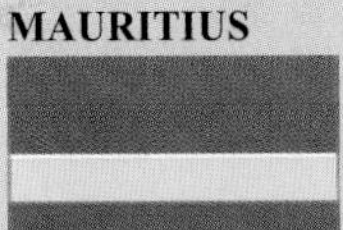
Pop: 1,230,602
Sq.Mil: 790
Sq.Km: 2,046

MOROCCO
Pop: 32,725,847
Sq.Mil: 172,414
Sq.Km: 446,550

MOZAMBIQUE

Pop: 19,406,703
Sq.Mil: 303,769
Sq.Km: 786,762

NAMIBIA

Pop: 2,030,695
Sq.Mil: 317,827
Sq.Km: 823,172

NIGER

Pop: 11,665,937
Sq.Mil: 489,189
Sq.Km: 1,267,000

NIGERIA

Pop: 128,771,988
Sq.Mil: 357,000
Sq.Km: 924,630

RWANDA

Pop: 8,440,820
Sq.Mil: 10,169
Sq.Km: 26,337

SAO TOME & PRINCIPE

Pop: 187,410
Sq.Mil: 372
Sq.Km: 963

SENEGAL

Pop: 11,126,832
Sq.Mil: 75,954
Sq.Km: 196,720

SEYCHELLES

Pop: 81,188
Sq.Mil: 176
Sq.Km: 455

SIERRA LEONE

Pop: 6,017,643
Sq.Mil: 27,925
Sq.Km: 72,325

SOMALIA

Pop: 8,591,629
Sq.Mil: 246,200
Sq.Km: 637,658

SOUTH AFRICA

Pop: 44,344,136
Sq.Mil: 455,318
Sq.Km: 1,179,274

SUDAN

Pop: 40,187,486
Sq.Mil: 967,494
Sq.Km: 2,505,809

SWAZILAND

Pop: 1,173,900
Sq.Mil: 6,705
Sq.Km: 17,366

TANZANIA
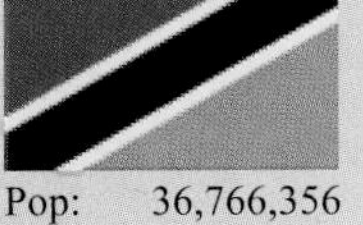
Pop: 36,766,356
Sq.Mil: 363,708
Sq.Km: 942,003

TOGO
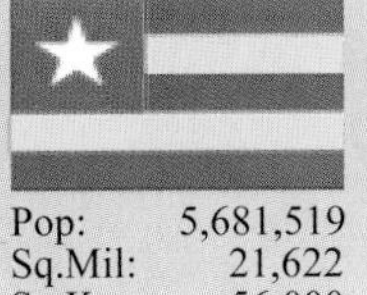

Pop: 5,681,519
Sq.Mil: 21,622
Sq.Km: 56,000

TUNISIA

Pop: 10,074,951
Sq.Mil: 63,378
Sq.Km: 164,149

UGANDA

Pop: 27,269,482
Sq.Mil: 91,076
Sq.Km: 235,887

ZAMBIA
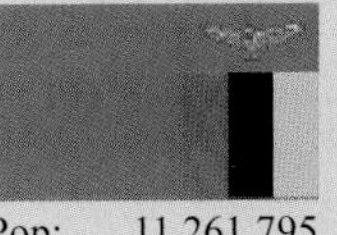
Pop: 11,261,795
Sq.Mil: 290,586
Sq.Km: 752,618

ZIMBABWE
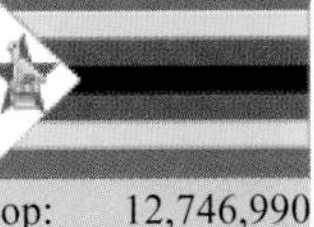
Pop: 12,746,990
Sq.Mil: 150,803
Sq.Km: 390,580

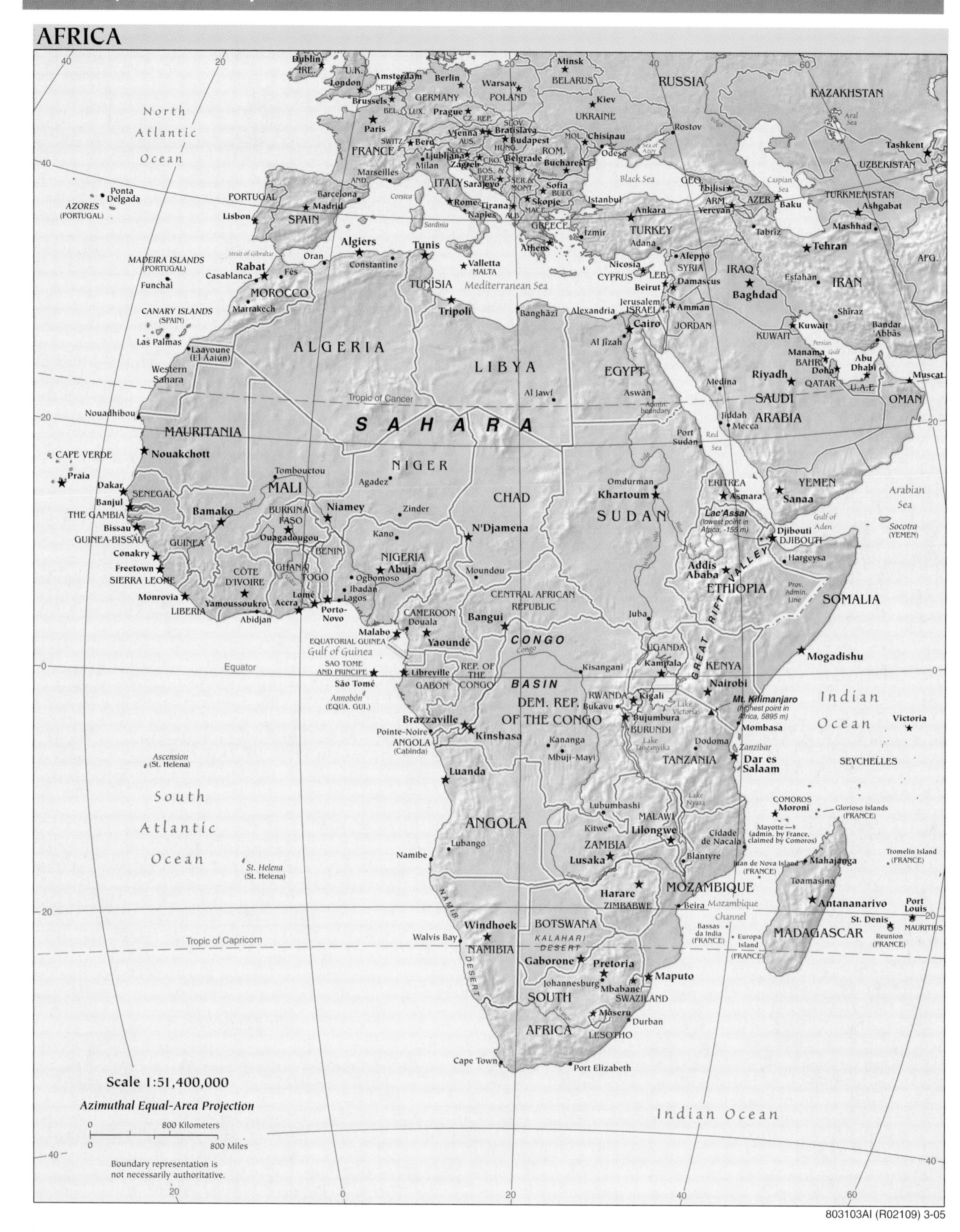

803103AI (R02109) 3-05

THE 47
INDEPENDENT NATIONS OF ASIA
FLAG, POPULATION &
LAND AREA
(SQ. MILES / SQ. KILOMETERS)

AFGHANISTAN

Pop: 29,928,987
Sq.Mil: 250,775
Sq.Km: 649,507

ARMENIA

Pop: 2,982,904
Sq.Mil: 11,506
Sq.Km: 29,800

AZERBAIJAN

Pop: 7,911,974
Sq.Mil: 33,436
Sq.Km: 86,600

BAHRAIN

Pop: 688,345
Sq.Mil: 240
Sq.Km: 622

BANGLADESH

Pop: 144,319,628
Sq.Mil: 55,126
Sq.Km: 142,776

BHUTAN

Pop: 2,232,291
Sq.Mil: 18,147
Sq.Km: 47,000

BRUNEI

Pop: 372,361
Sq.Mil: 2,226
Sq.Km: 5,765

BURMA/
MYANMAR

Pop: 42,909,464
Sq.Mil: 261,789
Sq.Km: 678,034

CAMBODIA/
KAMPUCHEA

Pop: 13,607,069
Sq.Mil: 69,898
Sq.Km: 181,036

CHINA

Pop: 1,306,313,812
Sq.Mil: 3,691,000
Sq.Km: 9,559,690

CYPRUS

Pop: 780,133
Sq.Mil: 3,473
Sq.Km: 8,995

EAST TIMOR

Pop: 1,040,880
Sq.Mil: 5,794
Sq.Km: 15,007

GEORGIA

Pop: 4,677,401
Sq.Mil: 26,911
Sq.Km: 69,700

INDIA

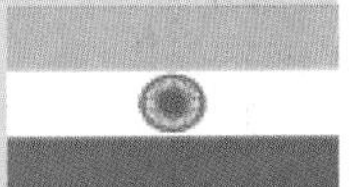

Pop: 1,080,264,388
Sq.Mil: 1,269,339
Sq.Km: 3,287,588

INDONESIA

Pop: 241,973,879
Sq.Mil: 788,430
Sq.Km: 2,042,034

IRAN

Pop: 68,017,860
Sq.Mil: 636,293
Sq.Km: 1,648,000

IRAQ

Pop: 26,074,906
Sq.Mil: 172,476
Sq.Km: 446,713

ISRAEL

Pop: 6,276,883
Sq.Mil: 7,847
Sq.Km: 20,324

JAPAN

Pop: 127,417,244
Sq.Mil: 145,730
Sq.Km: 377,441

JORDAN

Pop: 5,759,732
Sq.Mil: 35,000
Sq.Km: 96,650

KAZAKHSTAN

Pop: 15,185,844
Sq.Mil: 1,048,300
Sq.Km: 2,715,100

KOREA, NORTH

Pop: 22,912,177
Sq.Mil: 46,540
Sq.Km: 120,539

KOREA, SOUTH

Pop: 48,422,644
Sq.Mil: 38,175
Sq.Km: 98,873

KUWAIT

Pop: 2,335,648
Sq.Mil: 6,532
Sq.Km: 16,918

KYRGYZSTAN

Pop: 5,146,281
Sq.Mil: 76,641
Sq.Km: 198,500

LAOS

Pop: 6,217,141
Sq.Mil: 91,428
Sq.Km: 236,800

LEBANON

Pop: 3,826,018
Sq.Mil: 4,015
Sq.Km: 10,399

MALAYSIA

Pop: 23,953,136
Sq.Mil: 128,308
Sq.Km: 332,318

MALDIVES

Pop: 349,106
Sq.Mil: 116
Sq.Km: 300

MONGOLIA

Pop: 2,791,272
Sq.Mil: 606,163
Sq.Km: 1,569,962

NEPAL

Pop: 27,676,547
Sq.Mil: 54,663
Sq.Km: 141,577

OMAN

Pop: 3,001,583
Sq.Mil: 120,000
Sq.Km: 310,800

PAKISTAN

Pop: 162,419,946
Sq.Mil: 310,403
Sq.Km: 803,944

PHILIPPINES

Pop: 87,857,473
Sq.Mil: 115,707
Sq.Km: 299,681

QATAR

Pop: 863,051
Sq.Mil: 4,247
Sq.Km: 11,000

SAUDI ARABIA

Pop: 26,417,599
Sq.Mil: 829,995
Sq.Km: 2,149,687

SINGAPORE

Pop: 4,425,720
Sq.Mil: 226
Sq.Km: 585

SRI LANKA

Pop: 20,064,776
Sq.Mil: 25,332
Sq.Km: 65,610

SYRIA

Pop: 18,448,752
Sq.Mil: 71,498
Sq.Km: 185,180

TAJIKISTAN

Pop: 7,163,506
Sq.Mil: 55,251
Sq.Km: 143,100

THAILAND

Pop: 65,444,371
Sq.Mil: 198,455
Sq.Km: 513,998

TURKEY

Pop: 69,660,559
Sq.Mil: 300,946
Sq.Km: 779,450

TURKMENISTAN

Pop: 4,952,081
Sq.Mil: 188,455
Sq.Km: 488,100

UNITED ARAB EMIRATES

Pop: 2,563,212
Sq.Mil: 32,278
Sq.Km: 83,600

UZBEKISTAN

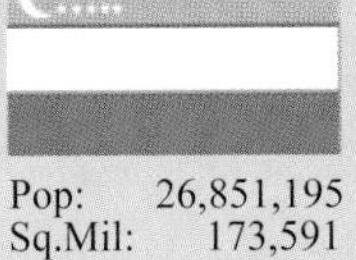

Pop: 26,851,195
Sq.Mil: 173,591
Sq.Km: 449,600

VIETNAM

Pop: 83,535,576
Sq.Mil: 128,405
Sq.Km: 332,569

YEMEN

Pop: 20,727,063
Sq.Mil: 188,321
Sq.Km: 487,752

ASIA

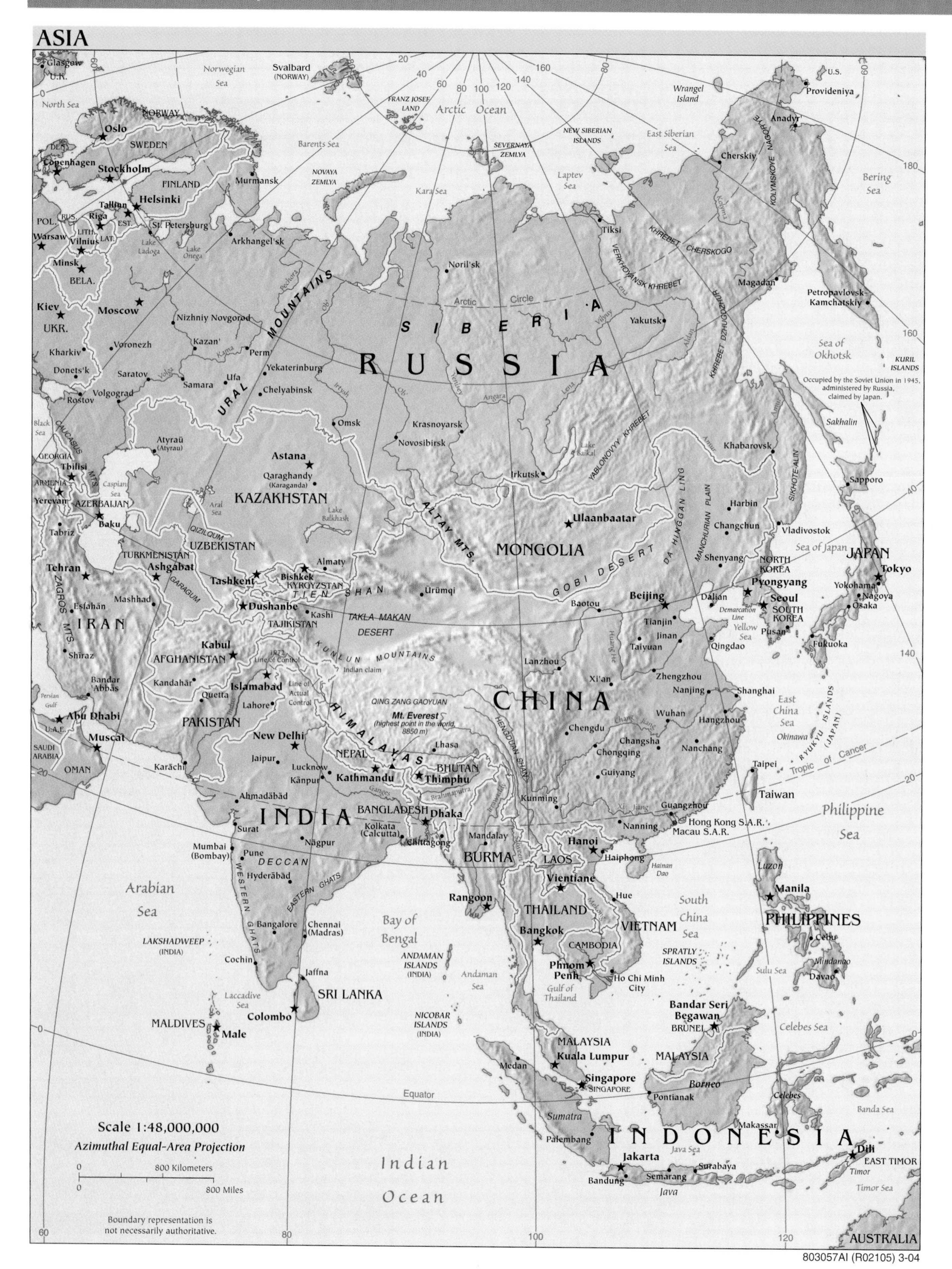

803057AI (R02105) 3-04

THE 43
INDEPENDENT NATIONS OF EUROPE
FLAG, POPULATION &
LAND AREA
(SQ. MILES / SQ. KILOMETERS)

ALBANIA

Pop: 3,563,112
Sq.Mil: 11,100
Sq.Km: 28,749

ANDORRA

Pop: 70,549
Sq.Mil: 188
Sq.Km: 487

AUSTRIA

Pop: 8,184,691
Sq.Mil: 32,375
Sq.Km: 83,851

BELARUS

Pop: 10,300,483
Sq.Mil: 80,154
Sq.Km: 207,600

BELGIUM

Pop: 10,364,388
Sq.Mil: 11,781
Sq.Km: 30,513

BOSNIA & HERZEGOVINA

Pop: 4,025,476
Sq.Mil: 19,940
Sq.Km: 51,645

BULGARIA

Pop: 7,450,349
Sq.Mil: 42,823
Sq.Km: 110,912

CROATIA

Pop: 4,495,904
Sq.Mil: 22,050
Sq.Km: 57,110

CZECH REPUBLIC

Pop: 10,241,138
Sq.Mil: 30,449
Sq.Km: 78,863

DENMARK

Pop: 5,432,335
Sq.Mil: 16,629
Sq.Km: 43,069

ESTONIA

Pop: 1,332,893
Sq.Mil: 17,413
Sq.Km: 45,100

FINLAND

Pop: 5,223,442
Sq.Mil: 130,128
Sq.Km: 337,032

FRANCE

Pop: 60,656,178
Sq.Mil: 210,038
Sq.Km: 543,998

GERMANY

Pop: 82,431,390
Sq.Mil: 137,753
Sq.Km: 356,780

GREECE

Pop: 10,668,354
Sq.Mil: 50,944
Sq.Km: 131,945

HUNGARY

Pop: 10,006,835
Sq.Mil: 35,919
Sq.Km: 93,030

ICELAND

Pop: 296,737
Sq.Mil: 39,768
Sq.Km: 103,000

IRELAND

Pop: 4,015,676
Sq.Mil: 27,136
Sq.Km: 70,282

ITALY

Pop: 58,103,033
Sq.Mil: 116,303
Sq.Km: 301,225

LATVIA

Pop: 2,290,237
Sq.Mil: 24,595
Sq.Km: 63,700

LIECHTEN-STEIN

Pop: 33,717
Sq.Mil: 62
Sq.Km: 161

LITHUANIA

Pop: 3,596,617
Sq.Mil: 25,174
Sq.Km: 65,200

LUXEMBOURG

Pop: 468,571
Sq.Mil: 999
Sq.Km: 2,587

MACEDONIA

Pop: 2,045,262
Sq.Mil: 9,889
Sq.Km: 25,612

MALTA

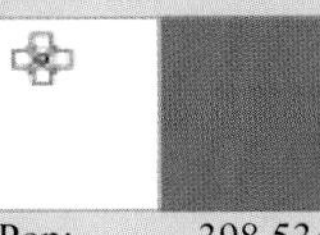

Pop: 398,534
Sq.Mil: 122
Sq.Km: 316

MOLDOVA

Pop: 4,455,421
Sq.Mil: 13,012
Sq.Km: 33,700

MONACO

Pop: 32,409
Sq.Mil: 0.75
Sq.Km: 2

THE NETHERLANDS

Pop: 16,407,491
Sq.Mil: 15,892
Sq.Km: 41,160

NORWAY

Pop: 4,593,041
Sq.Mil: 125,053
Sq.Km: 323,887

POLAND

Pop: 38,635,144
Sq.Mil: 120,725
Sq.Km: 312,678

PORTUGAL

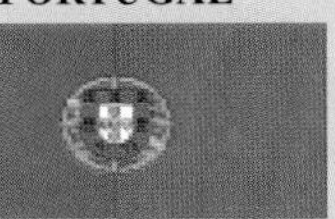

Pop: 10,566,212
Sq.Mil: 35,549
Sq.Km: 92,072

ROMANIA

Pop: 22,329,977
Sq.Mil: 91,699
Sq.Km: 237,500

RUSSIA

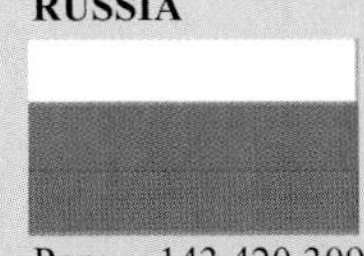

Pop: 143,420,309
Sq.Mil: 6,592,812
Sq.Km: 17,075,400

SAN MARINO

Pop: 28,880
Sq.Mil: 24
Sq.Km: 62

SERBIA & MONTENEGRO

Pop: 10,829,175
Sq.Mil: 39,517
Sq.Km: 102,350

SLOVAKIA

Pop: 5,431,363
Sq.Mil: 18,924
Sq.Km: 49,013

SLOVENIA

Pop: 2,011,070
Sq.Mil: 7,898
Sq.Km: 20,455

SPAIN

Pop: 40,341,462
Sq.Mil: 194,881
Sq.Km: 504,742

SWEDEN

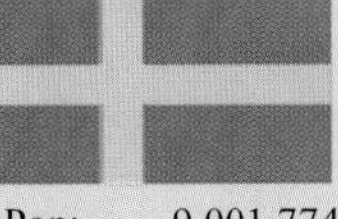

Pop: 9,001,774
Sq.Mil: 173,665
Sq.Km: 449,792

SWITZERLAND

Pop: 7,489,370
Sq.Mil: 15,943
Sq.Km: 41,292

UKRAINE

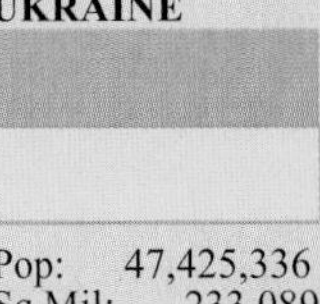

Pop: 47,425,336
Sq.Mil: 233,089
Sq.Km: 603,700

UNITED KINGDOM

Pop: 60,441,457
Sq.Mil: 94,399
Sq.Km: 244,493

VATICAN CITY /
HOLY SEE

Pop: 921
Sq.Mil: 0.17
Sq.Km: 0.44

EUROPE

Greenland (DENMARK)
Jan Mayen (NORWAY)
Barents Sea
Greenland Sea
Denmark Strait
Hammerfest
Murmansk
Tromsø
Kiruna
Norwegian Sea
White Sea
Arkhangel'sk
Reykjavík
ICELAND
Arctic Circle
Luleå
Oulu
NORWAY
Umeå
FINLAND
Lake Onega
Trondheim
Tórshavn
Faroe Islands (DENMARK)
SWEDEN
Gulf of Bothnia
Lake Ladoga
Tampere
SHETLAND ISLANDS
Turku
Helsinki
Gävle
St. Petersburg
Gulf of Finland
Bergen
ALAND ISLANDS
RUSSIA
Oslo
Stockholm
Tallinn
ESTONIA
Rockall (U.K.)
ORKNEY ISLANDS
HEBRIDES
Stavanger
Volga
Moscow
North Atlantic Ocean
Aberdeen
Skagerrak
Göteborg
Gotland
Riga
LATVIA
Kattegat
Glasgow
Edinburgh
North Sea
Baltic Sea
Öland
LITHUANIA
Vitsyebsk
Smolensk
Belfast
UNITED
DENMARK
Malmö
Vilnius
Mahilyow
Copenhagen
Kaliningrad
RUSSIA
Minsk
Isle of Man (U.K.)
Dublin
Irish Sea
Bornholm
BELARUS
Leeds
Manchester
Liverpool
Gdańsk
Hrodna
IRELAND
Homyel'
Hamburg
KINGDOM
Birmingham
Amsterdam
NETH.
Bremen
Berlin
Poznań
Warsaw
Brest
Cardiff
Rotterdam
Oder
POLAND
Kiev
London
Łódź
Rivne
Celtic Sea
Essen
Cologne
Leipzig
Elbe
Wrocław
Vistula
Dnieper
English Channel
Brussels
Lille
BEL.
Bonn
GERMANY
Prague
L'viv
UKRAINE
Guernsey (U.K.)
Jersey (U.K.)
Frankfurt am Main
Kraków
CZECH REPUBLIC
Chernivtsi
Luxembourg
SLOVAKIA
CARPATHIAN MTS.
Mykolayiv
Paris
LUX.
Brno
Bratislava
Chisinau
Seine
Stuttgart
Strasbourg
Rhine
Munich
Danube
Iași
Vienna
Odesa
Nantes
Loire
Budapest
Cluj-Napoca
MOLDOVA
LIECH.
AUSTRIA
Zürich
Vaduz
HUNGARY
Bay of Biscay
Bern
SWITZ.
FRANCE
Geneva
ALPS
ROMANIA
Ljubljana
SLOVENIA
Bucharest
Constanța
MASSIF CENTRAL
Lyon
Milan
Zagreb
Black Sea
Turin
Venice
Po
DINARIC ALPS
Bordeaux
BOSNIA AND HERZEGOVINA
Belgrade
Varna
Danube
Bilbao
Genoa
CROATIA
SAN MARINO
SERBIA AND MONTENEGRO
Toulouse
MONACO
Sarajevo
BULGARIA
Andorra la Vella
Marseille
Ligurian Sea
Florence
Sofia
Porto
PYRENEES
APENNINES
Adriatic Sea
Skopje
Istanbul
Zaragoza
ANDORRA
ITALY
MACEDONIA
Corsica
Rome
Madrid
Bursa
PORTUGAL
VATICAN CITY
Tirana
Thessaloniki
Barcelona
ALB.
Naples
TURKEY
Lisbon
Tagus
Balearic Sea
Sardinia
Tyrrhenian Sea
SPAIN
Valencia
Aegean Sea
İzmir
GREECE
BALEARIC ISLANDS
Athens
Sevilla
Cagliari
Ionian Sea
Mediterranean Sea
Palermo
Gibraltar (U.K.)
Málaga
Sicily
Rhodes
Strait of Gibraltar
Ceuta (SPAIN)
Alborán Sea
Algiers
Scale 1: 19,500,000
Crete
Lambert Conformal Conic Projection, standard parallels 40°N and 56°N
Melilla (SPAIN)
Oran
Tunis
Valletta
MALTA
300 Kilometers
Rabat
Casablanca
TUNISIA
300 Miles
MOROCCO
ALGERIA
Boundary representation is not necessarily authoritative.
803104AI (R01083) 3-05

THE 23
INDEPENDENT NATIONS OF NORTH AMERICA
FLAG, POPULATION & LAND AREA (SQ. MILES / SQ. KILOMETERS)

ANTIGUA & BARBUDA

Pop: 68,722
Sq.Mil: 171
Sq.KM: 443

BAHAMAS, THE

Pop: 301,790
Sq.Mil: 5,382
Sq.KM: 13,939

BARBADOS

Pop: 279,254
Sq.Mil: 166
Sq.KM: 430

BELIZE

Pop: 279,457
Sq.Mil: 8,867
Sq.KM: 22,966

CANADA

Pop: 32,805,041
Sq.Mil: 3,851,787
Sq.KM: 9,976,139

COSTA RICA

Pop: 4,016,173
Sq.Mil: 19,575
Sq.KM: 50,700

CUBA

Pop: 11,346,670
Sq.Mil: 44,206
Sq.KM: 114,494

DOMINICA

Pop: 69,029
Sq.Mil: 290
Sq.KM: 751

DOMINICAN REPUBLIC

Pop: 8,950,034
Sq.Mil: 18,704
Sq.KM: 48,443

EL SALVADOR

Pop: 6,704,932
Sq.Mil: 8,260
Sq.KM: 21,393

GRENADA

Pop: 89,502
Sq.Mil: 133
Sq.KM: 344

GUATEMALA

Pop: 14,655,189
Sq.Mil: 42,042
Sq.KM: 108,889

HAITI

Pop: 8,121,622
Sq.Mil: 10,694
Sq.KM: 27,697

HONDURAS

Pop: 6,975,204
Sq.Mil: 43,277
Sq.KM: 112,087

JAMAICA

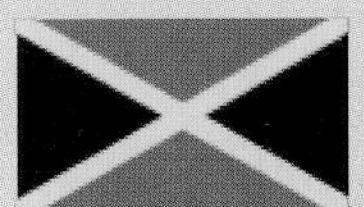

Pop: 2,731,832
Sq.Mil: 4,411
Sq.KM: 11,424

MEXICO

Pop: 106,202,903
Sq.Mil: 761,601
Sq.KM: 1,972,546

NICARAGUA

Pop: 5,465,100
Sq.Mil: 45,698
Sq.KM: 118,358

PANAMA

Pop: 3,039,150
Sq.Mil: 29,761
Sq.KM: 77,082

SAINT KITTS & NEVIS

Pop: 38,958
Sq.Mil: 103
Sq.KM: 269

SAINT LUCIA

Pop: 166,312
Sq.Mil: 238
Sq.KM: 616

SAINT VINCENT & THE GRENADINES

Pop: 117,534
Sq.Mil: 150
Sq.KM: 388

TRINIDAD & TOBAGO

Pop: 1,088,644
Sq.Mil: 1,980
Sq.KM: 5,128

UNITED STATES

Pop: 295,734,134
Sq.Mil: 3,623,420
Sq.KM: 9,384,658

NORTH AMERICA

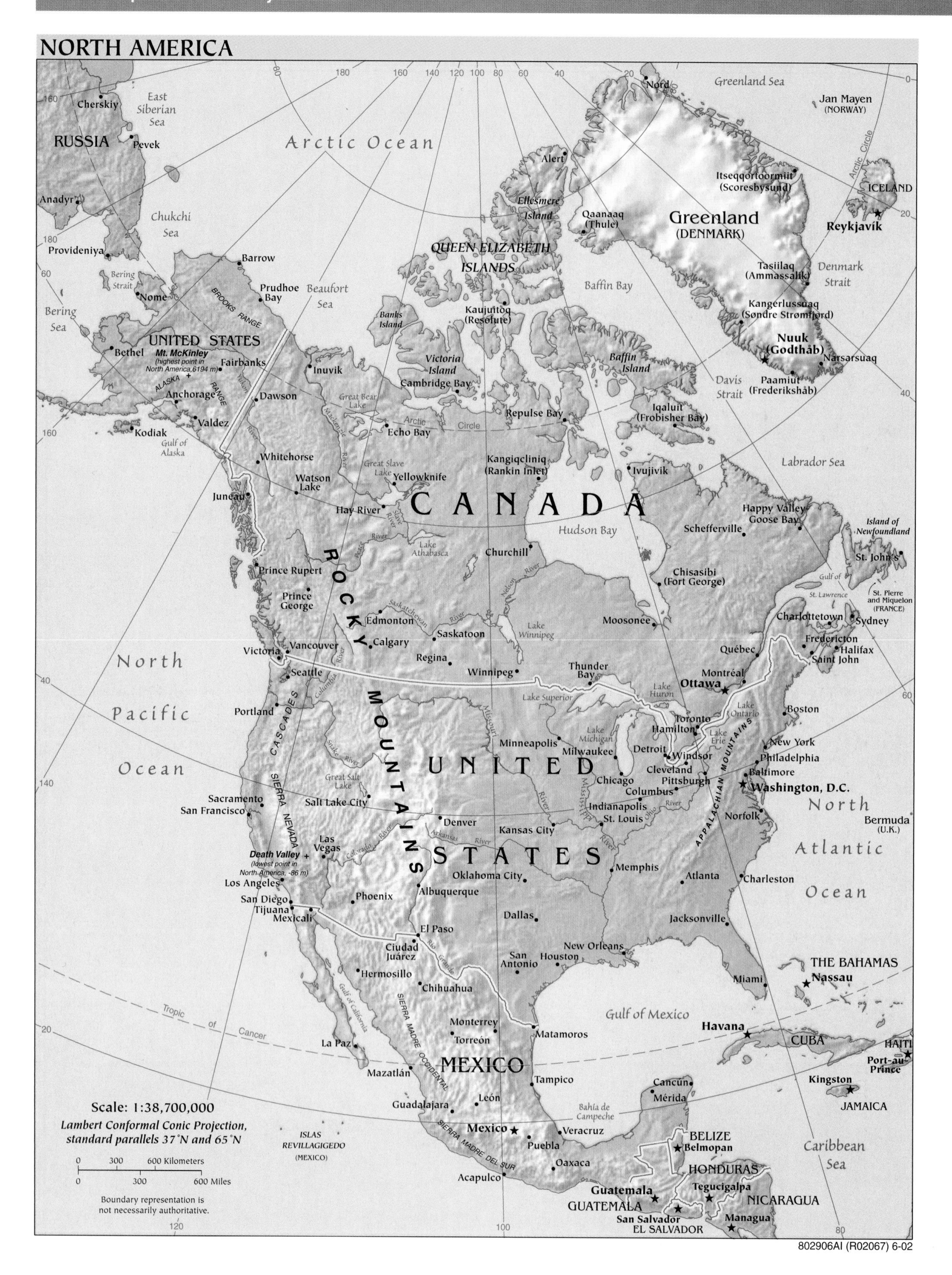

802906AI (R02067) 6-02

THE 12
INDEPENDENT NATIONS OF SOUTH AMERICA
FLAG, POPULATION &
LAND AREA
(SQ. MILES / SQ. KILOMETERS)

ARGENTINA

Pop: 39,537,943
SqMil: 1,072,070
SqKm: 2,776,661

BOLIVIA

Pop: 8,857,870
SqMil: 424,163
SqKm: 1,098,582

BRAZIL

Pop: 186,112,794
SqMil: 3,284,426
SqKm: 8,506,663

CHILE

Pop: 15,980,912
SqMil: 292,257
SqKm: 756,946

COLOMBIA

Pop: 42,954,279
SqMil: 439,513
SqKm: 1,138,339

ECUADOR

Pop: 13,363,593
SqMil: 109,483
SqKm: 283,561

GUYANA

Pop: 765,283
SqMil: 83,000
SqKm: 214,970

PARAGUAY

Pop: 6,347,884
SqMil: 157,047
SqKm: 406,752

PERU

Pop: 27,925,628
SqMil: 496,222
SqKm: 1,285,215

SURINAME

Pop: 438,144
SqMil: 55,144
SqKm: 142,823

URUGUAY

Pop: 3,415,920
SqMil: 72,172
SqKm: 186,925

VENEZUELA

Pop: 25,375,281
SqMil: 352,143
SqKm: 912,050

SOUTH AMERICA

THE 14
INDEPENDENT NATIONS OF OCEANIA
FLAG, POPULATION &
LAND AREA
(SQ. MILES / SQ. KILOMETERS)

AUSTRALIA

Pop: 20,090,437
Sq.Mil: 2,966,136
Sq.Km: 7,682,300

FIJI

Pop: 893,354
Sq.Mil: 7,055
Sq.Km: 18,272

KIRIBATI

Pop: 103,092
Sq.Mil: 291
Sq.Km: 754

MARSHALL ISLANDS

Pop: 59,071
Sq.Mil: 70
Sq.Km: 181

MICRONESIA

Pop: 108,105
Sq.Mil: 271
Sq.Km: 702

NAURU

Pop: 13,048
Sq.Mil: 8
Sq.Km: 21

NEW ZEALAND

Pop: 4,035,461
Sq.Mil: 103,736
Sq.Km: 268,676

PALAU

Pop: 20,303
Sq.Mil: 188
Sq.Km: 487

PAPUA NEW GUINEA

Pop: 5,545,268
Sq.Mil: 183,540
Sq.Km: 475,369

SAMOA/*WESTERN SAMOA*

Pop: 177,287
Sq.Mil: 1,133
Sq.Km: 2,934

SOLOMON ISLANDS

Pop: 538,032
Sq.Mil: 11,500
Sq.Km: 29,785

TONGA

Pop: 112,422
Sq.Mil: 270
Sq.Km: 699

TUVALU

Pop: 11,636
Sq.Mil: 10
Sq.Km: 26

VANUATU

Pop: 205,754
Sq.Mil: 5,700
Sq.Km: 14,763

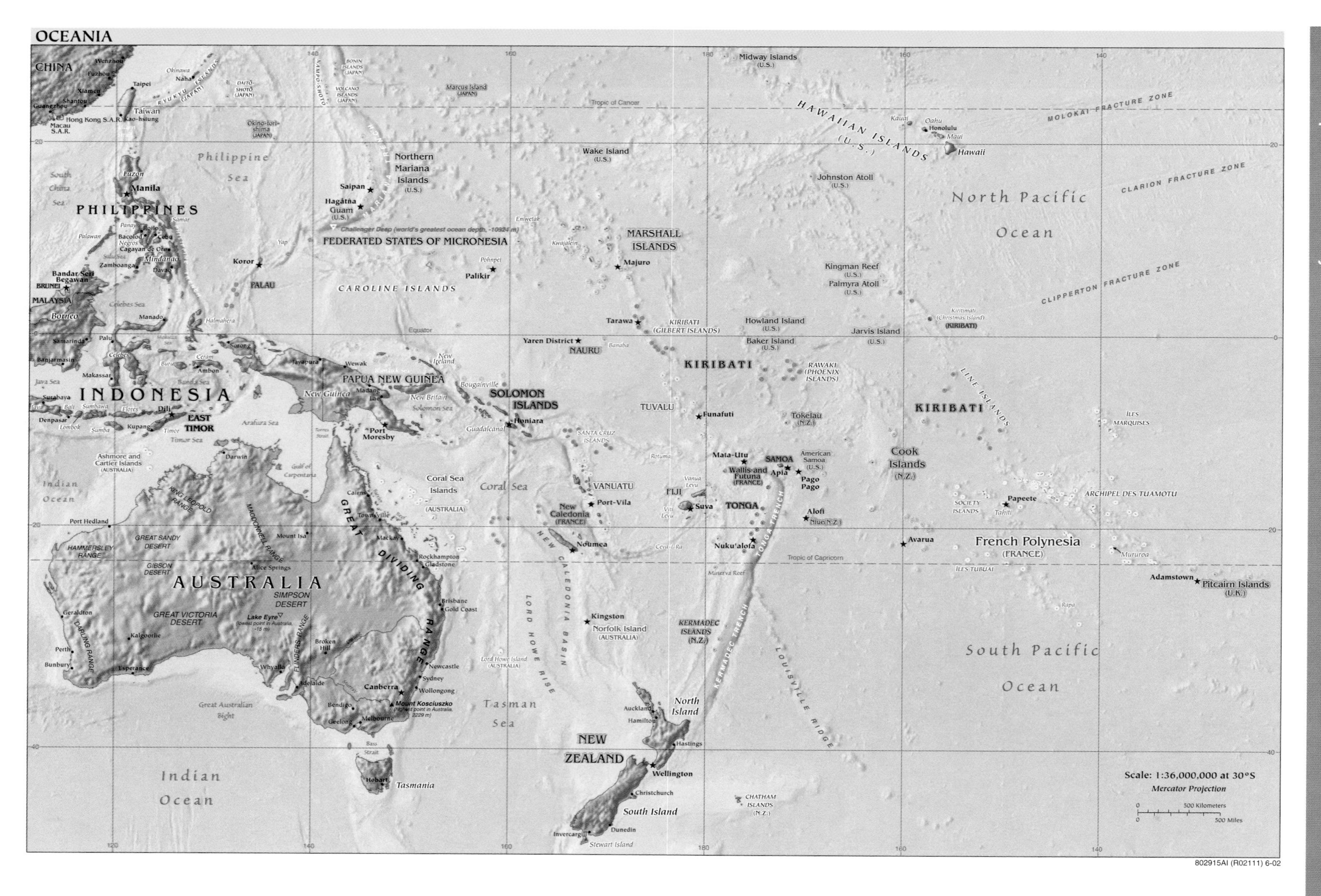
OCEANIA
CHINA
Taipei
Taiwan
Kao-hsiung
Hong Kong S.A.R.
Macau S.A.R.
Okinawa
Naha
Philippine Sea
PHILIPPINES
Manila
Luzon
Mindanao
Davao
Zamboanga
BRUNEI
Bandar Seri Begawan
MALAYSIA
Borneo
Celebes Sea
Koror
PALAU
Saipan
Hagåtña
Guam (U.S.)
Northern Mariana Islands (U.S.)
Challenger Deep (world's greatest ocean depth, -10924 m)
FEDERATED STATES OF MICRONESIA
Palikir
Pohnpei
CAROLINE ISLANDS
Marcus Island (JAPAN)
Wake Island (U.S.)
MARSHALL ISLANDS
Majuro
Tarawa
KIRIBATI (GILBERT ISLANDS)
Yaren District
NAURU
KIRIBATI
TUVALU
Funafuti
SOLOMON ISLANDS
Honiara
Guadalcanal
Bougainville
PAPUA NEW GUINEA
Port Moresby
Wewak
New Guinea
New Britain
INDONESIA
EAST TIMOR
Dili
Kupang
Denpasar
Surabaya
Makassar
Manado
Ambon
Java Sea
Arafura Sea
Timor Sea
Equator
Ashmore and Cartier Islands (AUSTRALIA)
Darwin
Gulf of Carpentaria
Cairns
Townsville
Mackay
Rockhampton
Gladstone
Brisbane
Gold Coast
Newcastle
Sydney
Wollongong
Canberra
Mount Kosciuszko
Melbourne
Geelong
Bendigo
Adelaide
Whyalla
Broken Hill
Mount Isa
Alice Springs
Port Hedland
Geraldton
Perth
Bunbury
Kalgoorlie
Esperance
AUSTRALIA
GREAT DIVIDING RANGE
GREAT SANDY DESERT
GIBSON DESERT
GREAT VICTORIA DESERT
SIMPSON DESERT
Lake Eyre
HAMMERSLEY RANGE
KING LEOPOLD RANGE
MACDONNELL RANGE
DARLING RANGE
FLINDERS RANGE
Great Australian Bight
Bass Strait
Hobart
Tasmania
Coral Sea Islands (AUSTRALIA)
Coral Sea
Tasman Sea
Indian Ocean
VANUATU
Port-Vila
New Caledonia (FRANCE)
Nouméa
NEW CALEDONIA BASIN
LORD HOWE RISE
Lord Howe Island (AUSTRALIA)
FIJI
Suva
TONGA
Nuku'alofa
TONGA TRENCH
KERMADEC TRENCH
KERMADEC ISLANDS (N.Z.)
LOUISVILLE RIDGE
SAMOA
Apia
American Samoa (U.S.)
Pago Pago
Wallis and Futuna (FRANCE)
Mata-Utu
Tokelau (N.Z.)
Niue (N.Z.)
Alofi
Cook Islands (N.Z.)
Avarua
KIRIBATI
LINE ISLANDS
French Polynesia (FRANCE)
Papeete
Tahiti
SOCIETY ISLANDS
ARCHIPEL DES TUAMOTU
ÎLES MARQUISES
ÎLES TUBUAI
Pitcairn Islands (U.K.)
Adamstown
NEW ZEALAND
Wellington
North Island
South Island
Auckland
Hamilton
Hastings
Christchurch
Dunedin
Invercargill
Stewart Island
CHATHAM ISLANDS (N.Z.)
Norfolk Island (AUSTRALIA)
Kingston
HAWAIIAN ISLANDS (U.S.)
Honolulu
Hawaii
Midway Islands (U.S.)
Johnston Atoll (U.S.)
Kingman Reef (U.S.)
Palmyra Atoll (U.S.)
Jarvis Island (U.S.)
Howland Island (U.S.)
Baker Island (U.S.)
RAWAKI (PHOENIX ISLANDS)
Kiritimati (Christmas Island) (KIRIBATI)
MOLOKAI FRACTURE ZONE
CLARION FRACTURE ZONE
CLIPPERTON FRACTURE ZONE
North Pacific Ocean
South Pacific Ocean
Tropic of Cancer
Tropic of Capricorn
Scale: 1:36,000,000 at 30°S
Mercator Projection
500 Kilometers
500 Miles
802915AI (R02111) 6-02

Dependencies and Territories of the World

COUNTRY / TERRITORY	***CAPITAL CITY***	**CLAIMED BY**	***CONTINENT***

American Samoa (*Pago Pago*) **USA** *(OC)*
Anguilla (*The Valley*) **Britain** *(NA)*
Aruba (*Oranjestad*) **The Netherlands** *(NA)*
Ashmore & Cartier Islands *(None)* **Australia** *(AS)*

Baker Island *(None)* **USA** *(OC)*
Bassas da India *(None)* **France** *(AF)*
Bermuda (*Hamilton*) **Britain** *(NA)*
Bouvet Island *(None)* **Norway** *(AN)*
British Indian Ocean Territory (*Diego Garcia*) **Britain** *(AS)*
British Virgin Islands (*Road Town*) **Britain** *(NA)*

Cayman Islands (*George Town*) **Britain** *(NA)*
Christmas Island (*Flying Fish Cove*) **Australia** *(AS)*
Clipperton Island *(None)* **France** *(NA)*
Cocos (Keeling) Islands (*West Island*) **Australia** *(AS)*
Cook Islands (*Avarua*) **New Zealand** *(OC)*
Coral Sea Islands *(None)* **Australia** *(OC)*

Europa Island *(None)* **France** *(AF)*

Falkland Islands (Stanley) **Britain** *(SA)*
Faroe Islands (Torshavn) **Denmark** *(EU)*
French Guiana (Cayenne) **France** *(SA)*
French Polynesia (*Papeete*) **France** *(OC)*
French Southern & Antarctic Lands *(None)* **France** *(AN)*

Gibraltar (*Gibraltar*) **Britain** *(EU)*
Glorioso Islands *(None)* **France** *(AF)*
Greenland (*Nuuk*) **Denmark** *(NA)*
Guadeloupe (*Basse-Terre*) **France** *(NA)*
Guam (*Agana*) **USA** *(OC)*
Guernsey (*St. Peter Port*) **Britain** *(EU)*

Heard & MacDonald Islands *(None)* **Australia** *(AN)*
Hong Kong (*None*) **China** *(AS)*
Howland Island *(None)* **USA** *(OC)*

Isle of Man (*Douglas*) **Britain** *(EU)*

Jan Mayen *(None)* **Norway** *(EU)*
Jarvis Island (*None*) **USA** *(OC)*
Jersey (*St. Helier*) **Britain** *(EU)*

Johnston Atoll (*None*) **USA** *(OC)*
Juan de Nova Island *(None)* **France** *(AF)*

Kingman Reef (*None*) **USA** *(OC)*

Macau (*Macau*) **China** *(AS)*
Martinique (*Fort-de-France*) **France** *(NA)*
Mayotte (*Mamoudzou*) **France** *(AF)*
Midway Islands (*None*) **USA** *(OC)*
Montserrat (*Plymouth*) **Britain** *(NA)*

Navassa Island (*None*) **USA** *(NA)*
Netherland Antilles (*Willemstad*) **The Netherlands** *(NA)*
New Caledonia (*Noumea*) **France** *(OC)*
Niue (*Alofi*) **New Zealand** *(OC)*
Norfolk Island (*Kingston*) **Australia** *(OC)*
Northern Mariana Islands (*Saipan*) **USA** *(OC)*

Palmyra Atoll (*None*) **USA** *(OC)*
Paracel Islands (*Woody Island*) **China** *(AS)*
Peter I Island (None) **Norway** *(EU)*
Pitcairn Islands (*Adamstown*) **Britain** *(OC)*
Puerto Rico (*San Juan*) **USA** *(NA)*

Queen Maud Land (None) **Norway** *(AN)*

Reunion (*Saint-Denis*) **France** *(AF)*

Saint Helena & Dependencies (*Jamestown*) **Britain** *(AF)*
Saint Pierre & Miquelon (*St. Pierre*) **France** *(NA)*
South Georgia & So. Sandwich Islands (*Grytviken*) **Britain** *(AN)*
Spratly Islands (*None*) **claimed by China, Vietnam &Others** *(AS)*
Svalbard *(Longyearbyen)* **Norway** *(EU)*

Taiwan/The Republic of China (*Taipei*) **China** *(AS)*
Tokelau (*None*) **New Zealand** *(OC)*
Tromelin Island *(None)* **France** *(AF)*
Turks & Caicos Islands (*Cockburn Town*) **Britain** *(NA)*

US Virgin Islands (*Charlotte Amalie*) **USA** *(NA)*

Wake Island (*None*) **USA** *(OC)*
Wallis & Futuna (*Mata Uta*) **France** *(OC)*

Africa *(AF)* **Antarctica** *(AN)* **Asia** *(AS)* **Europe** *(EU)* **North America** *(NA)* **South America** *(SA)* **Oceania** *(OC)*

The Arctic (North Pole) the Antarctic (South Pole) and Atlas

THE ARCTIC

This area is north of *The Arctic Circle*, with *The North Pole* right in the middle. It includes the Arctic Ocean, Greenland, Baffin Island, smaller northern islands, Norway and the far northern parts of Europe, Russia (Siberia), Alaska and Canada.

The Arctic Circle is an imaginary line located at 30° latitude, and 66° longitude, and it defines the southernmost part of the Arctic. The climate within the Circle is very cold and much of the area is always covered with ice.

During winter, the sun never rises and temperatures can reach in the lows of (- 50° F) in the higher latitudes. Further south, in summer, a 24 hours of sunlight a day melts the seas and topsoil, which causes the icebergs breaking off from the north and floating south, causing danger for the shipping routes of the northern Atlantic.

The residents of the Arctic include the Eskimos (Inuits), Lapps and Russians with a total population exceeding 2 million. The indigenous Eskimos have lived in the area for over 9,000 years, and many have now given up much of their traditional hunting and fishing to work in the oil fields and various villages.

The first explorers of the Arctic were Vikings. Then came the Norwegians who visited the northern regions in the 9th century. In 1909, after numerous attempts by regional explorers, Robert E. Peary reached the North Pole.

ARCTIC REGION

ANTARCTICA

Antarctica is the 7th continent and the 5th largest.

The Antarctic Circle and The South Pole at the center, it is almost completely covered with ice, and has active territorial claims submitted by Argentina, Australia, Chile, France, New Zealand, Norway and the United Kingdom.

(Since many of these claims are not recognized by some countries the continent remains in a constant disputed status).

Antarctica is surrounded by small islands and the following countries:

Falkland Islands (South America), and other smaller islands, Madagascar, South Africa, Tasmania, Australia, Chile, Argentina and New Zeland.

TIME ZONE

Here is a place where all time zones converge, therefore, everyone in Antarctica officially goes by New Zealand time.

CAPITAL CITY: None
OFFICIAL CURRENCY: None
OFFICIAL LANGUAGE: None
SIZE : 13,208,984 sq km (5,099,994 sq miles)

CLIMATE

Antarctica is the coldest and windiest spot on the planet. The lowest temperature ever recorded on Earth was recorded in Antarctica (-129.3°F) and the winter temperatures range from (-40° to -94°F). Normally winds are measured at about 200 miles per hour.

POPULATION

This is the only continent with no permanent population. However, there are governmental research stations with small groups of scientists from all over the world at all times. In addition an estimated 8,000 tourists visit the continent each year.

ANTARCTIC REGION

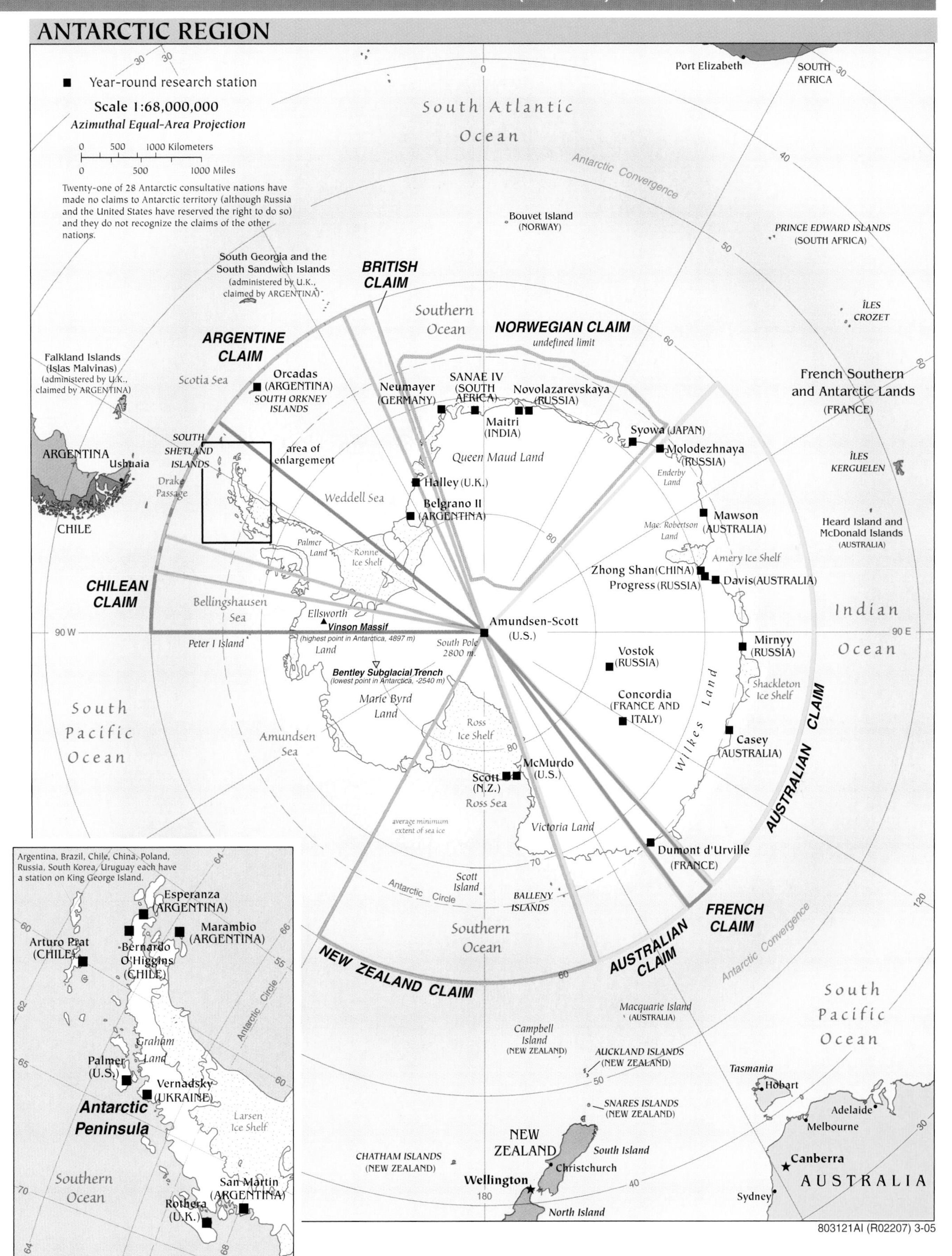

803121AI (R02207) 3-05

Physical Map of the World, April 2005
AUSTRALIA Independent state
Bermuda Dependency or area of special sovereignty
Sicily / AZORES Island / island group
Capital
CANADA
UNITED STATES
ROCKY MOUNTAINS
NORTH PACIFIC OCEAN
SOUTH PACIFIC OCEAN
NORTH ATLANTIC OCEAN
SOUTH ATLANTIC OCEAN
MID-ATLANTIC RIDGE
BRAZIL
ANDES
Greenland
SAHARA
ALGERIA
LIBYA
EGYPT
CHAD
NIGER
SUDAN
RUSSIA
SIBERIA
KAZAKHSTAN
CHINA
INDIA
INDIAN OCEAN
MID-INDIAN OCEAN RIDGE
AUSTRALIA
NEW ZEALAND
SOUTHERN OCEAN
Antarctica

The Original and Current Member States of the United Nations

♦ The Original 51 Member States in 1945

♦ Argentina ♦ Australia ♦ *Belarus (*)* ♦ Belgium ♦ Bolivia ♦ Brazil ♦ Canada ♦ Chile ♦ Chin ♦ Colombia ♦ Costa Rica ♦ Cuba ♦ *Czechoslovakia (*)* ♦ Denmark ♦ Dominican Republic ♦ Ecuador ♦ *Egypt (*)* ♦ El Salvador ♦ Ethiopia ♦ France ♦ Greece ♦ Guatemala ♦ Haiti ♦ Honduras ♦ India ♦ Iran ♦ Iraq ♦ Lebanon ♦ Liberia ♦ Luxembourg ♦ Mexico ♦ Netherlands, The ♦ New Zealand ♦ Nicaragua ♦ Norway ♦ Panama ♦ Paraguay ♦ Peru ♦ Philippines ♦ Poland ♦ *Russian Federation (*)* ♦ Saudi Arabia ♦ South Africa ♦ *Syrian Arab Republic (*)* ♦ Turkey ♦ Ukraine ♦ United Kingdom of Great Britain & Northern Ireland ♦ United States of America ♦ Uruguay ♦ Venezuela ♦ *Yugoslavia (*)*

The 140 Member States who joined the U.N. after 1945

ADMITED	MEMBER STATES		
1946	Afghanistan, Iceland, Sweden, Thailand	1962	Algeria, Burundi, Jamaica, Rwanda, Trinidad & Tobago, Uganda
1947	Pakistan, *Yemen (*)*		
1948	Myanmar/Burma	1963	Kenya, Kuwait
1949	Israel	1964	Malawi, Malta, Zambia
1950	*Indonesia (*)*	1965	Gambia, Maldives, Singapore
1955	Albania, Austria, Bulgaria, Cambodia, Finland, Hungary, Ireland, Italy, Jordan, Lao People's Democratic Republic, Libyan Arab Jamahiriya, Nepal, Portugal, Romania, Spain, Sri Lanka	1966	Barbados, Botswana, Guyana, Lesotho
		1967	Democratic Yemen
		1968	Equatorial Guinea, Mauritius, Swaziland
		1970	Fiji
		1971	Bahrain, Bhutan, Oman, Qatar, United Arab Emirates
1956	Japan, Morocco, Sudan, Tunisia	1973	Bahamas,The, *(*) Federal Republic of Germany, German Democratic Republic*
1957	Ghana, *Malaysia (*)*		
1958	Guinea		
1960	Benin, Burkina Faso, Cameroon, Central African Republic, Chad, Congo, Côte d'Ivoire/Ivory Coast, Cyprus, *Democratic Republic of the Congo(*)*, Gabon, Madagascar, Mali, Niger, Nigeria, Senegal, Somalia, Togo	1974	Bangladesh, Grenada, Guinea-Bissau
		1975	Cape Verde, Comoros, Mozambique, Papua New Guinea, Sao Tome & Principe, Suriname
		1976	Angola, Samoa, Seychelles
		1977	Djibouti, Viet Nam
1961	Mauritania, Mongolia, Sierra Leone, *United Republic of Tanzania (*)*	1978	Dominica, Solomon Islands
		1979	Saint Lucia

THE ORIGINAL FIFTY-ONE & CURRENT MEMBER STATES OF THE UNITED NATIONS

ADMITED	MEMBER STATES
1980	Saint Vincent & The Grenadines, Zimbabwe
1981	Antigua & Barbuda, Belize, Vanuatu
1983	Saint Kitts & Nevis
1984	Brunei Darussalam
1990	Liechtenstein, Namibia
1991	Democratic People's Republic of Korea, Estonia, Federated States of Micronesia, Latvia, Lithuania, Marshall Islands, Republic of Korea
1992	Armenia, Azerbaijan, Bosnia & Herzegovina, Croatia, Georgia, Kazakhstan, Kyrgyzstan, Moldova, San Marino, Slovenia,Tajikistan, Turkmenistan, Uzbekistan
1993	Andorra, Czech Republic, Eritrea, Monaco, Slovak Republic, The former Yugoslav Republic of Macedonia
1994	Palau
1999	Kiribati, Nauru, Tonga
2000	Tuvalu, Serbia & Montenegro
2002	**Switzerland, East Timor (Timor-Leste)**

TOTAL MEMBER STATES: 191
Please note, although there are 192 Independent Nations in the World, Vatican City/Holy See is not a member of the U.N.

NOTES

(*) BELARUS

ON 19 SEPTEMBER 1991, BYELORUSSIA INFORMED THE UNITED NATIONS THAT IT HAD CHANGED ITS NAME TO BELARUS.

(*) CZECH REPUBLIC / SLOVAKIA

Czechoslovakia was an original Member of the United Nations from 24 October 1945. In a letter dated 10 December 1992, its Permanent Representative informed the Secretary-General that the Czech and Slovak Federal Republic would cease to exist on 31 December 1992 and that the Czech Republic and the Slovak Republic, as successor States, would apply for membership in the United Nations. Following the receipt of its application, the Security Council, on 8 January 1993, recommended to the General Assembly that the Czech Republic be admitted to United Nations membership. The Czech Republic was thus admitted on 19 January of that year as a Member State.

(*) EGYPT / SYRIA

Egypt and Syria were original Members of the United Nations from 24 October 1945. Following a plebiscite on 21 February 1958, the United Arab Republic was established by a union of Egypt and Syria and continued as a single Member. On 13 October 1961, Syria, having resumed its status as an independent State, resumed its separate membership in the United Nations. On 2 September 1971, the United Arab Republic changed its name to the Arab Republic of Egypt.

(*) RUSSIA

The Union of Soviet Socialist Republics was an original Member of the United Nations from 24 October 1945. In a letter dated 24 December 1991, Boris Yeltsin, the President of the Russian Federation, informed the Secretary-General that the membership of the Soviet Union in the Security Council and all other United

THE ORIGINAL FIFTY-ONE & CURRENT MEMBER STATES OF THE UNITED NATIONS

Nations organs was being continued by the Russian Federation with the support of the 11 member countries of the Commonwealth of Independent States.

(*) YUGOSLAVIA

Bosnia & Herzegovina / Croatia / Macedonia / Serbia & Montenegro / Slovenia

The Socialist Federal Republic of Yugoslavia was an original Member of the United Nations, the Charter having been signed on its behalf on 26 June 1945 and ratified 19 October 1945, until its dissolution following the establishment and subsequent admission as ***new members of Bosnia and Herzegovina, the Republic of Croatia, the Republic of Slovenia, The former Yugoslav Republic of Macedonia, and The Federal Republic of Yugoslavia.***

The Republic of ***Bosnia & Herzegovina*** was admitted as a Member of the United Nations by General Assembly resolution A/RES/46/237 of 22 May 1992.

The Republic of ***Croatia*** was admitted as a Member of the United Nations by General Assembly resolution A/RES/46/238 of 22 May 1992.

By resolution A/RES/47/225 of 8 April 1993, the General Assembly decided to admit as a Member of the United Nations the State being provisionally referred to for all purposes within the United Nations as ***"The former Yugoslav Republic of Macedonia"*** pending settlement of the difference that had arisen over its name.

The Federal Republic of Yugoslavia was admitted as a Member of the United Nations by General Assembly resolution A/RES/55/12 of 1 November 2000.

Following the adoption and the promulgation of the Constitutional Charter of Serbia and Montenegro by the Assembly of the Federal Republic of Yugoslavia, on 4 February 2003, the name of the State of ***The Federal Republic of Yugoslavia was changed to Serbia & Montenegro.***

(*) YEMEN

Yemen was admitted to membership in the United Nations on 30 September 1947 and Democratic Yemen on 14 December 1967. On 22 May 1990, the two countries merged and have since been represented as one Member with the name "Yemen."

(*) INDONESIA

By letter of 20 January 1965, Indonesia announced its decision to withdraw from the United Nations "at this stage and under the present circumstances". By telegram of 19 September 1966, it announced its decision "to resume full cooperation with the United Nations and to resume participation in its activities". On 28 September 1966, the General Assembly took note of this decision and the President invited representatives of Indonesia to take seats in the Assembly.

(*) MALAYA

The Federation of Malaya joined the United Nations on 17 September 1957. On 16 September 1963, its name was changed to Malaysia, following the admission to the new federation of Singapore, Sabah (North Borneo) and Sarawak. Singapore became an independent State on 9 August 1965 and a Member of the United Nations on 21 September 1965.

(*) CONGO, DEMOCRATIC REPUBLIC OF THE

Zaire joined the United Nations on 20 September 1960. On 17 May 1997, its name was changed to Democratic Republic of the Congo.

(*) TANZANIA

Tanganyika was a Member of the United Nations from 14 December 1961 and Zanzibar was a Member from 16 December 1963. Following the ratification on 26 April 1964 of Articles of Union between Tanganyika and Zanzibar, the United Republic of Tanganyika and Zanzibar continued as a single Member, changing its name to the United Republic of Tanzania on 1 November 1964.

(*) GERMANY

The Federal Republic of Germany and the German Democratic Republic were admitted to membership in the United Nations on 18 September 1973. Through the accession of the German Democratic Republic to the Federal Republic of Germany, effective from 3 October 1990, the two German States have united to form one sovereign State.

The United States of America

THE UNITED STATES OF AMERICA'S STATES & CAPITAL CITIES

The Capital of The United States of America is Washington, District of Columbia (Washington, D.C.)

STATE		CAPITAL	STATE		CAPITAL
ALABAMA	(AL)	Montgomery	**MONTANA**	(MT)	Helena
ALASKA	(AK)	Juneau	**NEBRASKA**	(NE)	Lincoln
ARIZONA	(AZ)	Phoenix	**NEVADA**	(NV)	Carson City
ARKANSAS	(AR)	Little Rock	**NEW HAMPSHIRE**	(NH)	Concord
CALIFORNIA	(CA)	Sacramento	**NEW JERSEY**	(NJ)	Trenton
COLORADO	(CO)	Denver	**NEW MEXICO**	(NM)	Santa Fe
CONNECTICUT	(CT)	Hartford	**NEW YORK**	(NY)	Albany
DELAWARE	(DE)	Dover	**NORTH CAROLINA**	(NC)	Raleigh
FLORIDA	(FL)	Tallahassee	**NORTH DAKOTA**	(ND)	Bismarck
GEORGIA	(GA)	Atlanta	**OHIO**	(OH)	Columbus
HAWAII	(HI)	Honolulu	**OKLAHOMA**	(OK)	Oklahoma City
IDAHO	(ID)	Boise	**OREGON**	(OR)	Salem
ILLINOIS	(IL)	Springfield	**PENNSYLVANIA**	(PA)	Harrisburg
INDIANA	(IN)	Indianapolis	**RHODE ISLAND**	(RI)	Providence
IOWA	(IA)	Des Moines	**SOUTH CAROLINA**	(SC)	Columbia
KANSAS	(KS)	Topeka	**SOUTH DAKOTA**	(SD)	Pierre
KENTUCKY	(KY)	Frankfort	**TENNESSEE**	(TN)	Nashville
LOUISIANA	(LA)	Baton Rouge	**TEXAS**	(TX)	Austin
MAINE	(ME)	Augusta	**UTAH**	(UT)	Salt Lake City
MARYLAND	(MD)	Annapolis	**VERMONT**	(VT)	Montpelier
MASSACHUSETTS	(MA)	Boston	**VIRGINIA**	(VA)	Richmond
MICHIGAN	(MI)	Lansing	**WASHINGTON**	(WA)	Olympia
MINNESOTA	(MN)	St. Paul	**WEST VIRGINIA**	(WV)	Charleston
MISSISSIPPI	(MS)	Jackson	**WISCONSIN**	(WI)	Madison
MISSOURI	(MO)	Jefferson City	**WYOMING**	(WY)	Cheyenne

UNITED STATES

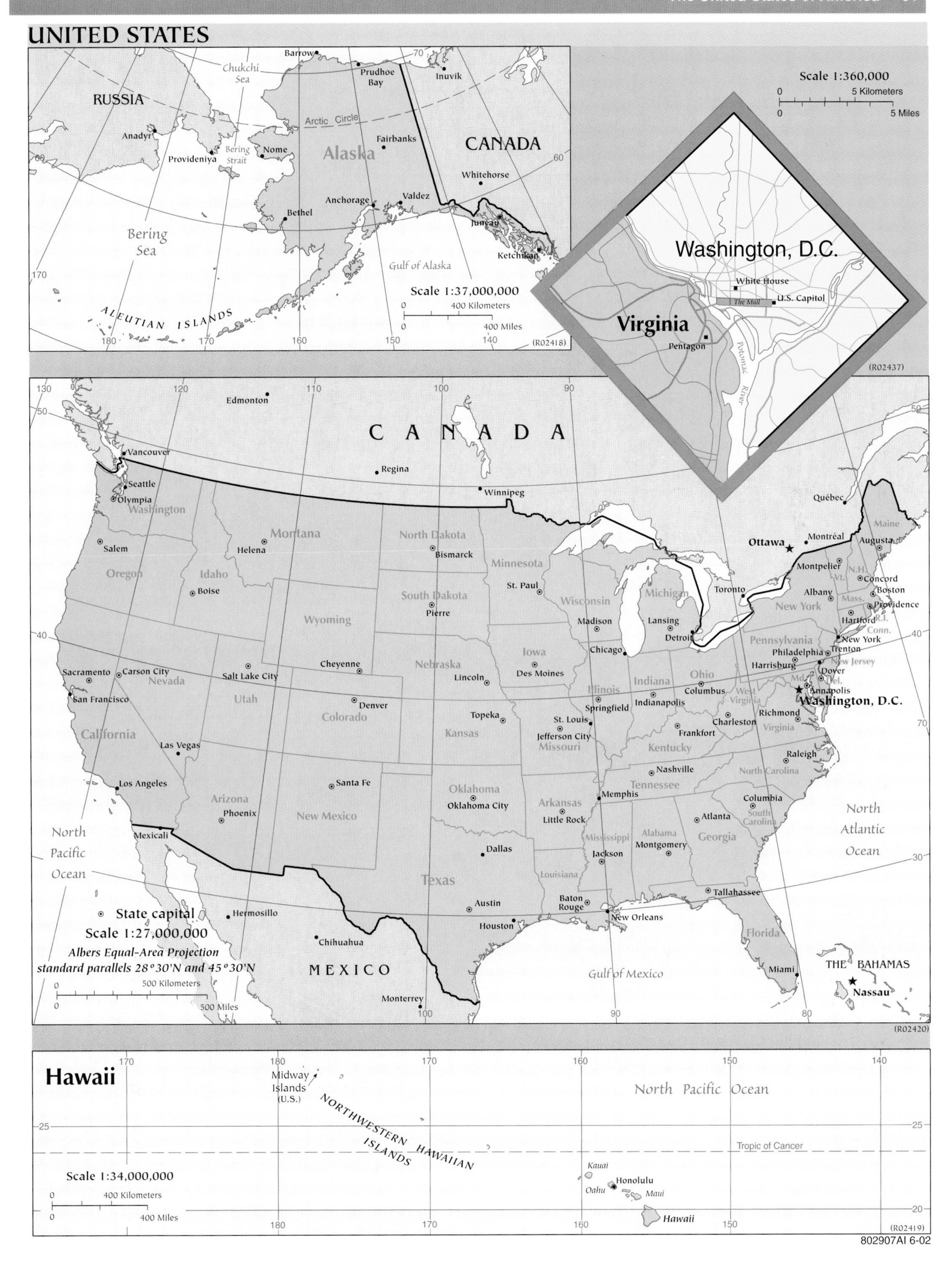

THE UNITED STATES OF AMERICA'S SYSTEM OF GOVERNMENT IN BRIEF

THE CONSTITUTION

Article I

THE LEGISLATIVE BRANCH
The Congress

2004 - 2005 Session / The 109th Congress

There are (2) Chambers
The Senate Chamber
The House of Representatives Chamber

Leadership Information ***(Both Chambers)***
Speakers of the House
Majority & Minority Leaders
Democratic Whips
Republican Whips
Democratic Caucus Chairmen &
Republican Conference Chairmen

The Senate

√ **100 Senators** *(Number stays the same)*
√ Divided into **Democrats & Republicans**
√ **2** Senators per State
√ **1** or more **Independent** *(On the outcome of election)*
√ **6** Years Term of Service
√ **1/3** of the seats are replaced every 2 years

The House of Representatives

√ **435 House of Representatives**
(The larger the State the moreRepresentatives_ Changes every 10 years, based on State population)
√ Divided into **Democrats & Republicans**
√ **1** or more **Independent** *(on outcome of election)*
√ **4 Delegates** *(Represent those outside the 50 States)*
√ **1 Resident Commissioner**
√ **2** Years Term of Service
√ **1/3** of the seats are replaced every 2 years

(GOP) stands for Grand Old Party

Article II

THE EXECUTIVES BRANCH
The White House

The President

The President is the Head of Government &
Chief of State
To become the President of The United States of America, one has to be born in the Country. The only Branch of Government with this requirement.

The President & Vice President

Elected on the same ticket by a college of
representatives;
who are elected directly from each State.
(In addition to the Popular votes,
out of a total of 538 Electrol votes,
270 votes are needed to win the Presidency).
Citizens of The US Territories
can not vote at the Presidential Election

The President & Vice President serve 4-years (a maximum of 2 terms of service__8 years)

Cabinet Members

Appointed by the President with Senate approval.
The Cabinet includes the Vice President and, by law,
the Heads of 15 Executive Departments
of the Secretaries

Secretary of ...

♦Agriculture ♦Commerce ♦Defense ♦Education ♦Energy ♦Health & Human Services ♦Homeland Security ♦Housing & Urban Development ♦Interior ♦Labor ♦State ♦Transportation ♦Treasury ♦Veterans Affairs **& The Department of Justice /The Attorney General**

Article III

THE JUDICIAL BRANCH
The Supreme Court

The Court has (9) Justices

(1) Chief Justice &
(8) Associate Justices

The justices are ***appointed for life***

The Justices are appointed by the President
with confirmation
by the Senate
United States Courts of Appeal
United States District Courts
State & County Courts

THE UNITED STATES OF AMERICA'S PRESIDENTS & VICE PRESIDENTS

TERM	PRESIDENT	PARTY	VICE PRESIDENT
1789-1797	1. George Washington	None	John Adams
1797-1801	2. John Adams	Federalist	Thomas Jefferson
1801-1809	3. Thomas Jefferson	Dem/Rep	Aaron Burr, George Clinton
1809-1817	4. James Madison	Dem/Rep	George Clinton, Elbridge Gerry
1817-1825	5. James Monroe	Dem/Rep	Daniel D. Tompkins
1825-1829	6. John Quincy Adams	Dem/Rep	John C. Calhoun
1829-1837	7. Andrew Jackson	Democratic	John C. Calhoun, Martin Van Buren
1837-1841	8. Martin Van Buren	Democratic	Richard M. Johnson
1841	9. William H. Harrison	Whig	John Tyler
1841-1845	10. John Tyler	Whig	None
1845-1849	11. James K. Polk	Democratic	George M. Dallas
1849-1850	12. Zachary Taylor	Whig	Millard Fillmore
1850-1853	13. Millard Fillmore	Whig	None
1853-1857	14. Franklin Pierce	Democratic	William R. King
1857-1861	15. James Buchanan	Democratic	John C. Breckinridge
1861-1865	16. Abraham Lincoln	Rep/Union	Hannibal Hamlin, Andrew Johnson
1865-1869	17. Andrew Johnson	Union	None
1869-1877	18. Ulysses S. Grant	Republican	Schuyler Colfax, Henry Wilson
1877-1881	19. Rutherford B. Hayes	Republican	William A. Wheeler
1881	20. James A. Garfield	Republican	Chester A. Arthur
1881-1885	21. Chester A. Arthur	Republican	None
1885-1889	22. Grover Cleveland	Democratic	Thomas A. Hendricks
1889-1893	23. Benjamin Harrison	Republican	Levi P. Morton
1893-1897	24. Grover Cleveland	Democratic	Adlai E. Stevenson
1897-1901	25. William McKinley	Republican	Garret A. Hobart, Theodore Roosevelt
1901-1909	26. Theodore Roosevelt	Republican	Charles W. Fairbanks
1909-1913	27. William H. Taft	Republican	James S. Sherman
1913-1921	28. Woodrow Wilson	Democratic	Thomas R. Marshall
1921-1923	29. Warren G. Harding	Republican	Calvin Coolidge
1923-1929	30. Calvin Coolidge	Republican	Charles G. Dawes
1929-1933	31. Herbert Hoover	Republican	Charles Curtis
1933-1945	32. Franklin D. Roosevelt	Democratic	John N. Garner, Henry A. Wallace, Harry S. Truman
1945-1953	33. Harry S. Truman	Democratic	Alben W. Barkley
1953-1961	34. Dwight D. Eisenhower	Republican	Richard M. Nixon
1961-1963	35. John F. Kennedy	Democratic	Lyndon B. Johnson
1963-1969	36. Lyndon B. Johnson	Democratic	Hubert H. Humphrey
1969-1974	37. Richard M. Nixon	Republican	Spiro T. Agnew, Gerald R. Ford
1974-1977	38. Gerald R. Ford	Republican	Nelson A. Rockefeller
1977-1981	39. Jimmy Carter	Democratic	Walter F. Mondale
1981-1989	40. Ronald W. Reagan	Republican	George H. W. Bush
1989-1993	41. George H. W. Bush	Republican	Dan Quayle
1993-2001	42. William J. Clinton	Democratic	Al Gore
2001-2009	43. George W. Bush	Republican	Richard B. Cheney

(1792) Irish-born Architect James Hoban designed The White House
(1801-1809) During President Jefferson's tenure, work on The White House was completed by architect Benjamin Latrobe
(1814) British troops burned The White House. Only the charred exterior walls were standing
(1817) The White House was rebuilt and reoccupied
The Office has been vacant a total of 37 years, because of 15 deaths & 3 resignations of Presidents and Vice Presidents

---

Aster Tessema
was born in Ethiopia.

She graduated from
Southeastern University,
Washington, D.C.,
with Business Administration Degree
and did post-graduate studies in
MBA, International Management.

She wrote the first Edition
of this book in 1982.

Aster began traveling
the world at a young age,
and visited numerous countries in
Africa, Asia, Europe,
North America & Oceania
